ELECTRONICS

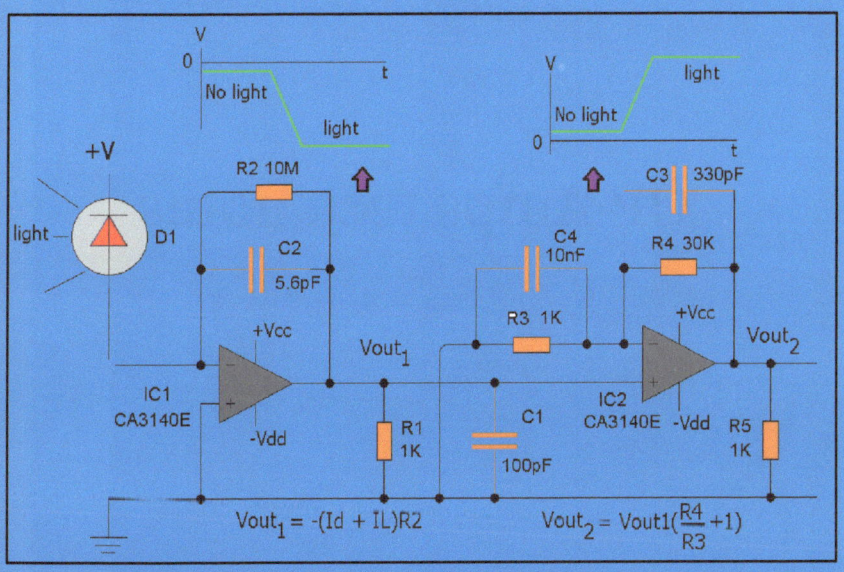

OP-AMPs tech notes

Dr. Fernando Moutinho

2019

ELECTRONICS

OP-Amps tech notes

Author: Dr. Fernando Moutinho

2019

With a lot of love to my wife Noemi and my son Kevin ....

Preface

Electronics is a very broad and extensive subject. In particular, because of there are many fields to cover. But fortunately, it can be grouped into two major groups. Analogic and Digital. However, in deep, everything is based in analogic electronics. In analogic electronics there are a lot of topics too. In a previous version the author wrote a book titled "Electrónica: Teoría y Aplicaciones prácticas de los Dispositivos más Comunes", Spanish version. In mentioned book is covered all the basic in electronics from the ohm's Law to the digital circuits, containing common devices and examples including theory and practice. Now, the author intends to make another contribution, this time concerning about: the OP-AMP.

There are many books and manuals about OP-AMPs covering many different topics and applications too. But this is a special book, written to understand the most common parameters that manufactures use and are detailed in the datasheet of each OP-AMP model. With an explanation about what is and the effect of each parameter in the OP-AMP behavior. So, the reader can have a more complete vision of the whole thing being able to use all the information from the manufacturer datasheet.

The learning activity is also accompanied by the applications of the OP-AMPs in basic circuits: amplifier, filter, summing, integrator, differentiator, comparator, etc. In each case the most important parameters are considered as well as good techniques to make a professional design. This is done by means of a lot of details in describing the circuit and by using simple mathematical demonstration.

As an additional effort the author proposes a very useful tables with summary of main formulas in each case and by selecting some of the most common OP-AMPs that can be used in almost all the applications, including recommended ones.

At the end of this book the author expects the reader can have a more complete understating about OP-AMPs parameters and how to use them, and consequently, to have a more powerful techniques to understand or making better and more professional designs.

About the author

Fernando Moutinho is Doctor and Magister in Electronics Instrumentation, graduated at the Universidad Central of Venezuela. He has more than 10 years of experience as professor and as researcher in surface physics.

With a lot of experience in physic and electronics at the University and in the industry, he had made interesting development in the scientific instrumentation related to: Mössbuer spectroscopy, radiation detectors, electronics microscopy, Auger spectroscopy, laser, data acquisition, signal conditioning, power supplies, process control, and more recently about IoT.

In concerning IoT field he is a referenced author in IoT courses on Udemy platform. But also, is very active with publication of books on the electronics subjects, IoT, and also with articles in recognized scientific journals.

The author has also a blog where he publishes some interesting information about topics in electronics, books, and IoT.

Table of Contents

Introduction 8

1. What is an OP-AMP? 10

2. OP-AMP Parameters 14

2.1 Open loop Gain (A_{OL}, A_{VOL}) 14

2.2 Output Impedance (Z_o, R_O) 16

2.3 Input Impedance (R_I, Z_{in}) 17

2.4 Frequency Response (Gain-Bandwidth Product) 17

2.5 Slew Rate (SR) 18

2.6 Total Harmonic Distortion (THD) 22

2.7 Equivalent input noise (e_{in}), ($\mathbf{nV\sqrt{Hz}}$) 25

2.8 Input Capacity (C_i) 31

2.9 CMRR (Common Mode Rejection Ratio) 35

2.10 Input Bias Current. 37

2.11 Input Offset Current (I_{OS}) 39

2.12 Input offset Voltage (V_{IO}, V_{OS}) 41

2.13 Output Current (I_O, OC) 41

2.14 Power Supply Rejection Ratio (PSRR) 42

3.0 Some practical circuits

3.1 Powering the OP-AMP 43

3.2 Inverter amplifier 46

3.3 Non-Inverter amplifier 56

3.4 Unity Gain (Buffer) 60

3.5 Inverter Sum Amplifier 62

3.6 Differential Amplifier 66

3.7 Inverter Integrator Amplifier 70

3.8 Charge Amplifier (photo detector) 75

3.10 Smith Trigger Comparator 79

4.0 OP-AMP Application table 83

Introduction

The present manuscript is intended to serve as a practical manual or tech guide for all those who love to design circuits with OP-AMPs. It is targeted to engineers, students, professors or amateurs, due to its useful technical information that is addressed with easy language and understandable mathematical expressions, also with examples. The content of this book is separated in 4 sections:

1. OP-AMP definition and how it works.
2. Description of the main parameters of the OPAMP and examples.
3. OP-AMP most common applications and their analysis.
4. A help-table to select an OP-AMP in applications.

In the section 1, the basic differential amplifier circuit is used to explain the base concept of the OP-AMP, and who it works. The explanation continues by mentioning its main features. In this section reader will understand very well what is inside of the OP-AMP, and how it works.

In the section 2, the most important parameters of the OP-AMP manufacture are descripted to understand what they are meaning, and their practical use in the designer role as well as the analysis. All selected parameters here are related to their corresponding in the manufacturer's datasheet.

Section 3 is about the practical use of the OP-AMP in the most common applications like: amplifying, summing, differentiation, integration, photo-detection and as comparator. This section is very detailed in explanation, calculation, and practical tips, to ensure a professional design and understanding. At the end of each design case a table containing a summary of all formulas and tips is found.

Finally, in the section 4, it has been elaborated an all in one-page help-table containing a summary. The idea is to have a fast look of most important parameters to select an OPAMP and its recommended applications. This pretends to help in the design by reducing time consumption looking at specific literature or heavy manuals. The help-table contents some of the most common OP-AMPs used in electronics.

This page in intentionally left in blank

1. What is an OP-AMP?

The acronym OP-AMP stands for **Op**erational **Amp**lifier, and it is well known type of amplifier which main characteristic is based on the use of the differential amplifier as the main input stage, followed by successive coupled stages in different configurations like: common emitter, common base, common collector or its equivalent configuration if JFET, CMOS or bipolar transistors are used.

Basically, the OP-AMP can perform operations like sum, differentiation, integration or derivation, from here his name of operational.

The coupling between differential amplifier and the rest of stages are generally made in DC. So, OP-AMP can work in both DC and AC mode.

At the input stage (first stage), the differential amplifier may be built using different types of transistors: BJT, JFET, CMOS or by a combination of all of them. The output of the OP-AMP is always equal to the difference between two inputs V_1 and V_2, that's why it is named differential. Then, this differential can be amplified by a gain parameter named here: $|A_V|$

The most relevant features of the OP-AMPs are:

1. Output is a difference between two inputs.
2. Has two mode of operation: differential and common
3. In the differential mode gain must be ≥ 1
4. In common mode the amplifier must exhibits high rejection or attenuation gain $<<1$.
5. The ratio between differential gain and the mode gain must be as high as possible (CMRR).

The mentioned ratio is known as Common Mode Rejection Ratio (CMRR) and it is expressed in dB units, and it is considered as the figure of merit in the qualification of the OP-AMP.

The figure 1 shows a simplified schematic of the differential amplifier. Note that at this moment is explaining about differential amplifier.

In reference to figure 1 we must consider the following:

Let's suppose now that we have two arbitrary input signals: V_1, V_2

By definition, the output of the differential amplifier will be: $v_{out} = A_v(V_2 - V_1)$

Where the term A_v represents here the overall gain of the amplifier being in differential or in common mode. Others terms to consider here and that are also present in the figure 1 are:

Z_{in} as the entrance impedance of the amplifier as seen by the input.

Z_{out} as the output impedance of the amplifier as seen by the load.

All These terms will be discussed in detail later.

Look at the figures 1 and 2 for the examples of differential and single amplifier mode configuration of the differential amplifier.

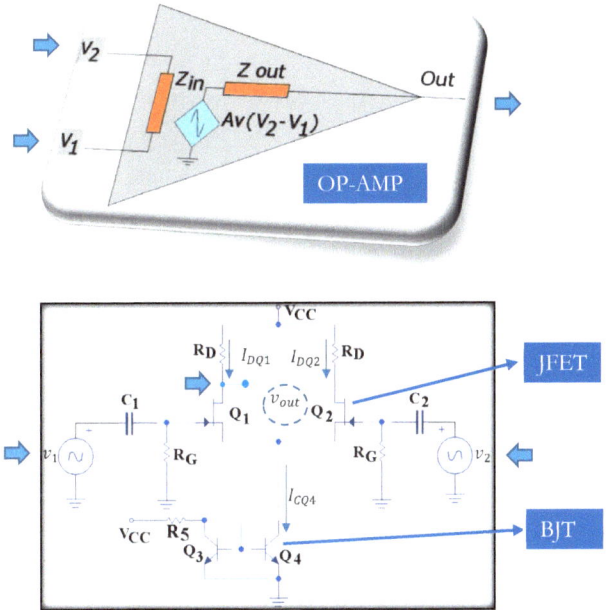

Figure 1. Differential amplifier: (a) block representation, (b) equivalent schematic.

To make thing more clear, the differential amplifier is a configuration that can be operated in dual or single mode, it is with 2 inputs or one input.

As for example V_1 and V_2 can be DC or AC signal, but most of the time we will refer to AC signal, so in the figure we will use capital letters but, in our analysis, we will use lowercase in reference to specifically AC signal.

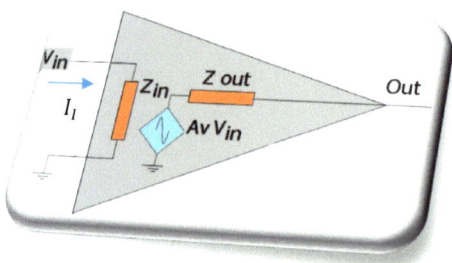

Figure 2. Simplified Schematic of the single amplifier.

The figure 1(a) shows the differential mode, as ports representation model, while figure 1(b) shows as example, a simplified internal schematic of a differential amplifier made with JFET and BJT type transistors, the ports for inputs and output are also indicated for comparison. This is just to have an idea about internal electronics. Figure 2 depicts the amplifier in simple mode.

Note that in single mode the amplifier is like it has grounded one of the inputs in the differential amplifier. Then, $V_1 = 0$, and V_2 signal becomes as a single V_{in}

Thus, the output of the single amplifier can be re-written as: $V_{out} = A_v(V_{in})$

Now let's suppose that we arrange a type of cascade configuration where differential amplifier is placed at the beginning followed by another single amplifier. Figure 3 shows such configuration.

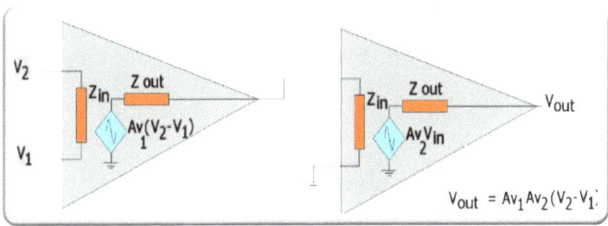

Figure 3. Cascade amplifier configuration: differential and a single stage coupled.

The output in this cascade amplifier is: $V_{out} = A_{v1} A_{v2} (V_2 - V_1)$

Note that the overall gain is now: $A_{v1} A_{v2}$, which means that whole gain of the amplifier is bigger but maintaining the operation in differential mode. Essentially, what we have done is a type of Operational Amplifier (OP-AMP) with only two stages.

And so how it works, in real live, OP-AMP may have many stages beginning with one or two differential amplifiers, and the rest is to multiply the gain, and one final stage with usually unitary gain for driving low impedance. About the final stage it will be discussed later.

The overall gain in the OP-AMPs can reach the order between 10^3 to 10^5 usually. So, one can imagine there are many stages multiplying in between. But the efficiency is in achieving this goal with the least amount of stages as possible, while maintaining the best parameters and the highest CMRR as possible. Then, some special configurations are made for this purpose. The purpose of this very high gain is convenient and will be noticed later.

There are many others parameters that make OP-AMPs great and ideal to use in electronics. In the next section these main parameters of the OP-AMPs will be discussed.

All the information concerning the parameters and characteristics of the OP-AMP is provided by the manufacturer in a pdf formatted document named: Datasheet, a standard document in which is rated all the information to use the OP-AMP in most applications. This Datasheet is commonly available through Internet. It is a good practice to find and check the datasheet prior or during the designing stage. There is a vast variety of OP-AMPs suitable for different applications.

2. OP-AMP Parameters

2.1 Open loop Gain (A_{OL}, A_{VOL})

The open Loop Gain can be defined as the maximum gain that amplifier can achieve without any feedback. It is denoted as A_{OL} or A_{VOL}. This parameter is also known as large signal voltage Gain. Numerically, can be expressed in decimal units, as for example 10^4, or in units of kV/V, as for example 100 kV/V, or in units of decibels (dB), as for example $20\log(1 \times 10^4) = 100$ dB.

The A_{OL} parameter is frequency dependent. When frequency increases until certain point (cut frequency $= f_c$) the A_{OL} decreases as a function of frequency until 0 dB gain ($A_{OL}=1$). Figure 4 shows an example of the frequency dependency of the A_{OL}.

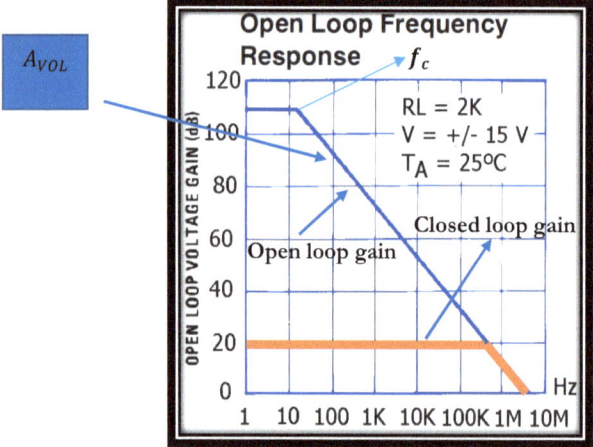

Figure 4. Frequency dependency of A_{OL}.

In the figure 4, the behavior of the A_{VOL} as function of the frequency is clearly shown. But also, the closed loop gain is shown, indicating the convenient of having a high open loop gain to let the close loop gain to have more room in frequency.

Note that there is a roll off frequency in which an inflection in the curve occurs. This point is calling the open loop cut frequency (f_c). Later, it will be demonstrated that product of gain by frequency is a constant parameter (Gain-Bandwidth product).

In the graph of the figure 4 the $A_{OL} \sim 110$ dB or ~ 315.000 for $f = 1$ Hz. As for example, for a frequency application about of 1MHz this OP-AMP with no feedback, will reduce the A_{OL} parameter about ~ 10 dB or lower. The corresponding gain will be about of 3 or lower. So, the effective gain will be much lower than A_{OL}. That why it is very important to know the frequency operation of the application. But, also the A_{OL} and bandwidth of the OP-AMP.

In the same case before, trying of using in a closed loop gain would be even worse, due to the negative feedback would tend to reduce even more the effective gain.

In terms of the ideal amplifier, the higher the A_{OL} the better the amplifier is. But real things are far to be ideals, so a compromise must be done to satisfy all requirements.

In rare occasion the OP-AMP will be used directly in an open loop mode, especially in the case of amplifying a signal. This is because of the excessive gain will turn output distorted and uncontrollable. That is why a negative feedback is used to reduce and to control the gain to desired level. As it was shown, a low gain has more bandwidth.

Negative closed loop is referred in any case where the output is feed to V^- port, of the OP-AMP, generally by a resistor network. It is a well-known and common technique used in OP-AMP for amplifying signals. There are several advantages in negative feedback, being one of them the extension of the frequency operation while the feedback gain (effective gain) is reduced. This effect will be understood when GBP parameter is discussed later.

2.2 Output Impedance (Z_o, R_O)

The output impedance (Z_o) is normally expressed in terms of the resistive component (R_o). Thus, assuming no reactive component. Ideally, this parameter must be zero (0 Ω) ohms. Its maximum value (R_O) is measured or calculated in the open loop mode and in DC, $f = 0\ Hz$.

Practically, it is in the order of 30-200 ohms (R_o). It is important to highlight that this impedance can be referred in both: open or closed loop mode. It is known that in a negative feedback the effective output impedance can be lowered several orders down.

In fact, while the closed loop gain is lower ($A_V \rightarrow 1$), lower the effective output impedance will be, and while the closed loop gain goes higher ($A_V \rightarrow \infty$), the effective output impedance goes higher too, to a maximum value of R_0, but in general tends to be lower than R_o. R_o is only temperature dependent.

In the case where the reactive component is included, output impedance is both frequency and temperature dependent. The output impedance is now called: Z_o.

An example of the typical behavior of the Z_O as a function of the closed loop gain and frequency is shown in figure 5. In the graph of the figure 5 is noticed that Z_0 varies with both: the effective gain (negative closed loop) and the frequency.

In the practice, most of the time this parameter is not critical, as long as the load (R_L) resistor keeps always much higher than effective R_0, then the output voltage is not reduced. Maximum load ($R_L=$ min) is calculated from the maximum output current the OP-AMP can drive, and in general R_L results always to be much higher than R_0.

Output impedance is very important because allows to determine the output reduction level caused by the internal R_0. Also, because, any voltage attenuated or dropped by R_0 will cause power dissipation, therefore, heating the entire device and possibly changing in some way the most of the parameters in the OP-AMP, especially input bias current, offset voltage, gain, and noise.

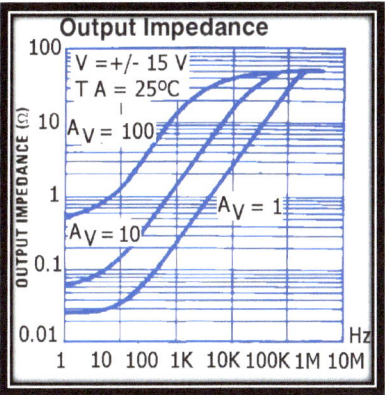

Figure 5. Output impedance R_0 as a function of frequency.

2.3 Input Impedance (R_I, Z_{in})

This parameter is one the closest to the ideal value. Input impedance of the OP-AMP could be in the range 10^3 to 10^{12} ohms, depending on what type of transistor the input is made; if BJT, JFET or CMOS, and its intended application. In most of the MOS or BiMOS type OP-AMPs input impedance can have very high values (1TΩ = $10^{12}\Omega$), which can be considered in practical terms as infinite. Bipolar type (BJT) has the lower impedance, in general hundreds of thousands of ohms (kΩ).

Very high impedance is always desired because an ideal amplifier must have a zero-input current. In the figure 2 input current at the amplifier is defined as:

$$I_I = \frac{V_{in}}{Z_{in}} \tag{1}$$

2.4 Frequency Response (Gain-Bandwidth Product)

Frequency response is expressed in terms of the Gain-Bandwidth Product (GBP). As it was mentioned before (2.1), this is a constant value parameter.

Figure 6 shows the GBP parameter as a function of the frequency. To build this curve (gain in decimal units) data has been tabulated from interpolation from the figure 4. Then, the product of the gain by frequency was obtained. As result the GBP obtained = 3.16 MHz.

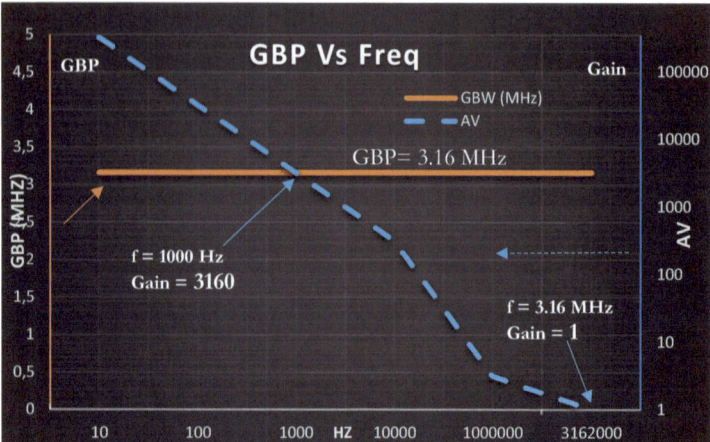

Figure 6. GBP parameter.

A solid straight line indicates that GBP is a constant all over the frequencies range of the OP-AMP. Thus, once GBP is known, is easy to calculate the maximum frequency or gain in which the OP-AMP can works without deterioration in gain or frequency.

As for example: in the figure 6 to work with 1kHz, the maximum gain to impose is about 3000. That means, if the frequency increases the real maximum gain will go down to keep constant GBP. Alternatively, if the frequency is lower the maximum gain obtainable will go higher. In other words, if a long bandwidth is required, a low gain is necessary. That is why the maximum frequency operation is reached with a gain of 1.

In the opposite way, if a high gain is desired, a narrower bandwidth is necessary.

2.5 Slew Rate (SR)

The slew rate is the ratio that measures how fast the output can go up or down, caused by a step change in the input. It is given in units of volts per time. The higher the slew rate is, the faster the output changes in the OP-AMP. Datasheet SR parameter normally represents its maximum value according to certain test conditions.

Typically, SR is expressed in unit of volt by microsecond (V/μs).

In the case where the output signal rate is slower than the input rate, the output signal is distorted exponentially. Thus, to avoid or minimize such problem a relation between frequency, height of the output, and the SR must be taking into a count.

Figure 7 is showing the effect of the SR in the output distortion.

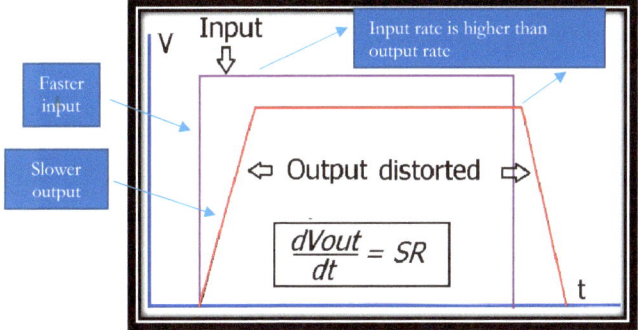

Figure 7. SR parameter.

As it is depicted in the figure 7 the SR can be defined as:

$$SR = \frac{dV_{out}}{dt} \qquad (2)$$

The SR is a limitation from OP-AMP design. But, in most of the time this limitation comes from an external component, like the output capacitor, that limits itself the rate that output changes. In this case, the output current has direct effect over the SR. Let's see:

$$V_{out} = \frac{1}{C} \int I_{out}\, dt \qquad (3)$$

Where:

I_{out} is the OP-AMP output current
V_{out} = output voltage
C = output capacitor

Then:
$$\frac{dV_{out}}{dt} = SR = \frac{I_{out}}{C} \qquad (4)$$

According to the equation (4), basically, the SR can be increased by pushing more current into the output capacitor.

The above equation is particularly useful when output is coupled to a stray capacitor or a smoothing filter (low pass) to the output signal port. In this case is recommendable to put some parallel resistor in the output to increase the output current, in such way that this capacitor will charge faster when current is higher, keeping the SR as high as possible. Of course, capacitor must be also as low as possible as indicated in the eq (4).

Figure 8 is a sample circuit which use a small capacitor (100 pF) to smooth the output in parallel with 1 kΩ resistor. The resistor guarantees enough current in the output to charge the capacitor faster enough. Then, the SR is keeping optimized by the external component.

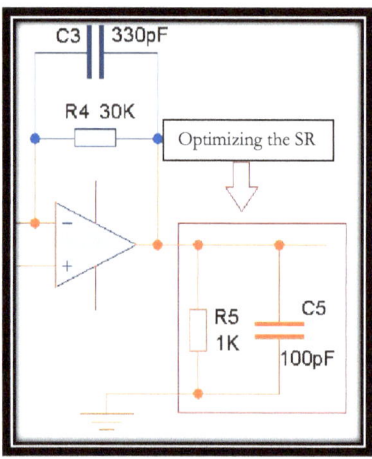

Figure 8. Optimizing SR parameter.

The limit for the resistor is the maximum current available for the OP-AMP.

As practical example if: V_{out} = 15 V, then I_{out} = 15 mA, and C = 100 pF

$$\frac{dV_{out}}{dt} = SR = \frac{I_{out}}{C} = 150 \, V/\mu s$$

If the manufacture given SR is 9V/µs, means that your circuit is not limiting this parameter, but is limited by OP-AMP, so SR in the circuit will be 9V/µs as it is the maximum value of the OP-AMP.

Careful must be taken not to put a load resistor below maximum allowed by datasheet. The output will be cropped (distorted) and can cause permanent damage in the OP-AMP.

The figure below is showing a case in which the SR parameter is lowered by the external circuit.

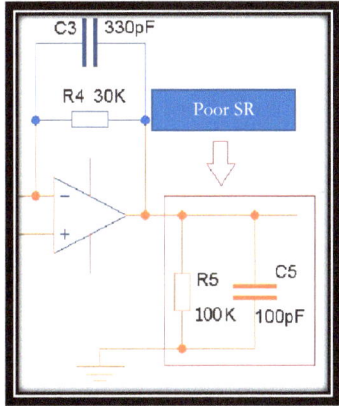

Figure 9. SR parameter.

$$\frac{dV_{out}}{dt} = SR = \frac{I_{out}}{C} = 1,5 \ V/\mu s$$

In the case of figure 9 the SR parameter will be 1.5 V/µs instead of 9v/µs. The effective SR is decreased. Therefore, distortion will be added at the output if input signal rate is higher than 1.5 V/µs. Distortion can be seen as any change in the output form.

When using this configuration make sure the cut frequency of the low pass filter formed by R_5 and C_5 is higher than your limit frequency of operation. Otherwise it will work as low pass filter. Thus, affecting the output in form and amplitude.

Another expression that can be used is shown in the equation (5). This equation is based on a sinusoidal input signal and determines de SR required to handle the signal at the output with minimal distortion.

$$SR = 2.\pi.f.V_{out} \quad \text{(sinusoidal)} \quad (5)$$

Where:

f= frequency of operation

As example, assuming same values from the case before, f= 10 kHz, V_{out} =15 V, and applying equation (5):

$$SR = 2.\pi.f.V_{out} = 0.94\ V/\mu s$$

The previous result shows that for 10 kHz sinusoidal input signal, the output SR required is 0.94 V/μs. If the OP-AMP gives a rate of 9V/μs, then output signal can follow the input steps very well, means as fast as the input. So, no distortion is introduced by SR.

As another example, assuming now 3 MHz frequency operation:

$$SR = 2.\pi.f.V_{out} = 282.7\ V/\mu s$$

In this case, the speed required is higher than the nominal value of 9V/μs, therefore, distortion will be present at the output, deforming the signal exponentially.

Sometimes if not possible to increase the SR, the distortion can be reduced a little by considering limiting or decreasing the output level. Thus, allowing the output to reach a certain level lower than the maximum, with a reasonable edges delay at the output. This technique is a reasonable compromise, if not output can be totally distorted.

2.6 Total Harmonic Distortion (THD)

Harmonics are the presence of odd and even multiples of the fundamental frequency (2*fo*, 3*fo*, 4*fo*, 5*f0*...) in the output of the amplifier due to the nonlinear response to an input frequency. The input frequency here is called the fundamental one (*fo*).

The total harmonic distortion (THD) is the ratio between the root mean square (RMS) of the sum of all harmonics contributions voltages divided by the RMS voltage of the fundamental frequency.

Mathematically is defined according to the equation (6):

$$THD = \frac{\sqrt{V_1^2 + V_2^2 + V_3^2 \ldots}}{V_0} \qquad (6)$$

Where V_0 is RMS value of the amplitude of the fundamental frequency f_0, and V_1, V_2, V_3 are the corresponding RMS amplitudes of the second, third and fourth… and so on successive considered harmonics of the: $2f_0$, $3f_0$, $4f_0$….

THD is usually expressed in percent unit.

Typical THD values in the OP-AMPs are about 0.05% or lower.

Figure 10 is an example of FFT (Fast Fourier Transform) spectrum corresponding to the output signal in a hypothetically amplifier, showing only the first 5 harmonics of the fundamental.

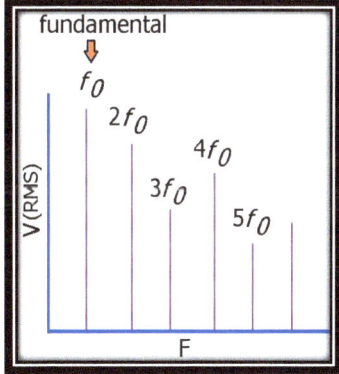

Figure 10. FFT spectrum example.

In the spectrum of figure 10 the amplitudes are measured in RMS voltages.

A sinusoidal signal can be used as fundamental frequency to test the amplifier THD, because the sinusoidal has no harmonics.

THD equation can also be expressed in terms of watts:

$$THD = \sqrt{\frac{P_1 + P_2 + P_3}{P_0}} \qquad (7)$$

Where:

P_0 is the power in watts of the fundamental frequency, and $P_1 \ldots P_n$ the power in watts of the rest of the harmonics.

Usually, the power spectrum is expressed in dBm or dBmW, both terminologies are interchangeable. The dBm is unit of power relative to one milliwatt (mW). Therefore, 0 dBm = 1 mW.

$$dBm = 10\log(mW)$$

To convert dBm in mW use the equation (8)

$$dBm\ to\ mW = e^{0.230258 * dBm} \qquad (mW) \qquad (8)$$

As example 70 dBm = 10.000.000 mW or 10.000 Watts = 10 kW.

To convert dBm in Watts use the equation (8.1)

$$dBm\ to\ W = \frac{e^{0.230258 * dBm}}{10^3} \qquad (W) \qquad (8.1)$$

To convert dBm in kW use the equation (8.2)

$$dBm\ to\ kW = \frac{e^{0.230258 * dBm}}{10^6} \qquad (kW) \qquad (8.2)$$

For example, spectrum of the output signal taken from the oscilloscope gives the results shown in the table 1:

THD can be calculated from values in table 1.

Freq (kHz)	Harmonic	dBm	P (mW)
1 (fundamental)	0	21.7	147.9
2	1	3.5	2.38
3	2	-6	0.25
4	3	-20	0.01
5	4	-23	0.005
6	5	-40	0.0001

Table 1. THD calculation example.

Then:

$$THD = \sqrt{\frac{2.38 + 0.25 + 0.01 + 0.005 + 0.0001}{147.9}} = 0.1337$$

$$THD\% = 13.37\ \%$$

If oscilloscope provides unit of Peak voltage, RMS value can be obtained dividing the peak value per square root of two ($\sqrt{2}$).

$$RMS = \frac{V_{paek}}{\sqrt{2}} \qquad (9)$$

Then, THD can obtained using equation (6).

2.7 Equivalent input noise (e_{in}), ($nV/\sqrt{Hz}$)

The equivalent input noise (e_{in}) is the total output noise divided by the open loop gain (A_{VOL}).

The total output noise of the amplifier is in any case the sum of all noise's contributions.

The equation (10) gives the expression for its calculation:

$$e_{TRMS} = \sqrt{e_{n1RMS}^2 + e_{n2RMS}^2 + e_{n3RMS}^2..}$$

$$e_{in} = \frac{e_{TRMS}}{A_{VOL}} \qquad (10)$$

Noise can come from different sources, as for example:

- Thermal Noise (Johnson noise)
- Shot noise (Schottky noise)
- $1/f$ (flicker or pink noise) noise.

Thermal noise: It is related to the thermal motion of electrons in the conductor. It is temperature and frequency dependent. It is associated to resistors. Mathematically expressed by the equation (11).

$$e_{thRMS} = \sqrt{4KTR\ \Delta f} \qquad (11)$$

Where:

K = Boltzmann constant = 1.38×10^{-23} Jk^{-1}

T = temperature in Kelvin. Example: 25°C = 298.15 K. T (K) = °C + 273.15

Δf = Bandwidth in Hz.

For example: at 25°C, 1kHz bandwidth a 100 MΩ resistor will generate a noise voltage of:

$$e_{th} = \sqrt{4KTR\, \Delta f} = \sqrt{1.645 10^{-9}} = 4.06 10^{-5}\ V \sim 40\ \mu V\ RMS$$

According to this, it is convenient not to use excessive high values resistors when is not absolutely necessary, because it will introduce more noise than necessary in the system.

Shot Noise: It is related to imperfections in the semiconductor device and associated to the flow of current across it. Charges does not arrive all at the same time from one electrode to another, thus, generating a statistical fluctuation in the current. As mentioned, this fluctuation has statistical behavior and can be calculated using equation (12). Shot noise is independent of the temperature. Equation (12) shows that.

$$e_{shRMS} = \sqrt{2QI\Delta f} \quad\quad (12)$$

Where:

Q = electron charge constant = 1.602×10^{-19} Coulomb.s

I = Average DC current in the semiconductor.

For example, for 1 KHz bandwidth, a photodiode with a dark current = 30nA, will generate a shot noise of:

$$e_{sh} = \sqrt{2QI\Delta f} = \sqrt{9.612^{-24}} \sim 3^{-12} pA$$

The above result seems to be very small, but to measure current about femtoamp for example, this noise will be very high and makes not possible to do it. In such case would be recommendable to down the operation current as much as possible, let´s say to less than 0.1 pA, and even also downing the frequency.

Contrary, in the case to measure current about mA range this noise is completely negligible.

$1/f$ noise: Its origin is not very clear yet, but it is present in passive and active components. It decreases as the frequency increases hence it is called *1/f*. It is associated with DC current.

Since it is not very well defined is better to say that reducing the DC current consumption of the circuit a lower the 1/f noise will be associated.

The $1/f$ noise spectrum has its larger contribution down to 1 kHz.

Total noise in OP-AMP is expressed in terms of $nV/\sqrt{Hz}$. The noise spectrum has a negative logarithmic decay. That means noise is decreasing when frequency gets higher. Figure 11 shows an example of the equivalent input noise as a function of the frequency.

Note that in figure 11 the maximum noise is about ~40 $nV/\sqrt{Hz}$ at 10 Hz. Most of the OP-AMPs manufactures report values around this in the datasheet.

Two expression are used: equivalent wideband input noise voltage, referred as maximum total noise in wide range of frequencies, and equivalent input noise referred for specific frequency, typically, 1 kHz or 10 kHz as reference frequencies.

According to this in the figure 11 will be:

- Equivalent Wideband input noise voltage ~ 40 nV maximum, BW = 10 Hz – 100 kHz
- Equivalent input noise voltage ~ 7.5 nV, f = 1 k Hz
- Equivalent input noise voltage ~ 3.5 nV, f = 10 k Hz

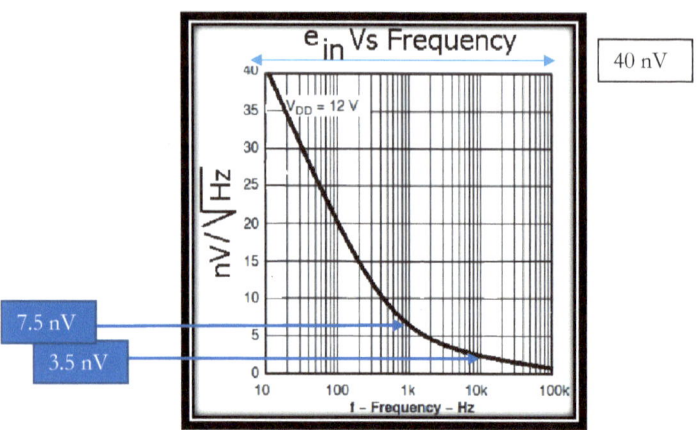

Figure 11. OP-AMP Equivalent input noise example.

The biggest contribution of noise in the equivalent input noise (figure 11) is $1/f$. It is supposed that thermal noise contribution and shot noise are not a significant amount in the total noise spectrum, especially for frequencies below 1 kHz.

Notice that noise is decreasing when frequency increasing, even when the term $\sqrt{Hz}$ suggests it is proportionally in all the noise equations. But the fact is the output voltage is decreasing as low pass filter, due to capacitor effects. Thus, the equivalent input noise will be also smaller since output is smaller too.

Figure 12 shows same spectrum in full logarithmic scale

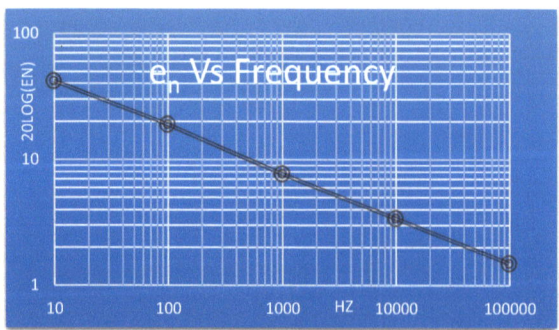

Figure 12. Equivalent input noise $e_n = 20\log (nV/\sqrt{Hz})$

Figure 12 clearly shows the logarithmic decay of total the input noise in the OP-AMP.

Figure 13 is shown a photodiode amplifier circuit used as example to calculate the total output noise from contributions:

According to what was mentioned before, the noise contributions considered here are:

-Shot noise from the reverse current of photodiode D_1.

-Thermal noise from resistors R_2, R_3.

-Equivalent input noise from the OP-AMP datasheet.

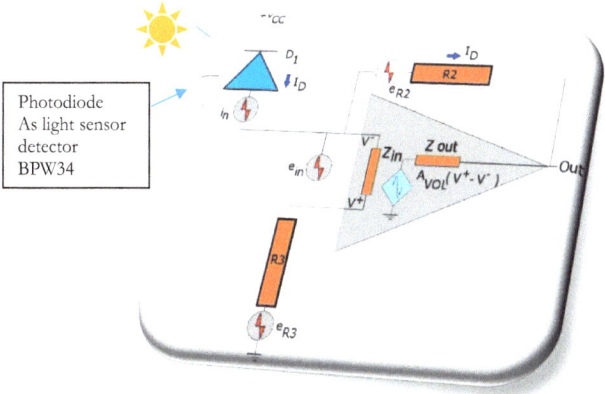

Figure 13. Noise in the photodiode amplifier.

In the circuit of figure 13, the OP-AMP output noise equation can be written like this:

$$V_{nout} = \sqrt{(I_n R_2)^2 + e_{R3}^2 + (e_{in} A_{VOL})^2 + e_{R2}^2}$$

Where:

$I_n = \sqrt{2QI_D}$; contribution noise from photodiode current I_D

$e_{R3} = \sqrt{4KTR_3}$; contribution noise from Thermal from R_3

$e_{in} = 12\ nV\ ;\ from\ OP-AMP\ datasheet, 1\ KHz\ (assumed)$;

$e_{R2} = \sqrt{4KTR_2}$; contribution noise from Thermal from R_2

Electronics: OP-AMP tech notes...! | *Equivalent input noise (ein), ($nV\sqrt{Hz}$)*

Assuming numerical examples values:

T = 25°C, R_2 = 10 MΩ, R_3 = 100 kΩ, I_D = 10nA, A_{VOL} = 315.000:

$I_n = \sqrt{2QI_D} = 5.65 \times 10^{-14}$ A

$e_{R3} = \sqrt{4KTR_3} = 4.06 \times 10^{-8}$ V

$e_{in} = 12$ nV

$e_{R2} = \sqrt{4KTR_2} = 4.06 \times 10^{-7}$ V

Substituting:

$$V_{nout} = \sqrt{(5.65 \times 10^{-14} \times 10^6)^2 + (4.06 \times 10^{-8})^2 + (3.78 \times 10^{-3})^2 + (4.06 \times 10^{-7})^2}$$

$$V_{nOUT} = \sqrt{3.19 \times 10^{-15} + 1.64 \times 10^{-15} + 1.42 \times 10^{-5} + 1.64 \times 10^{-13}}$$

$$V_{nOUT} = 3.76 \times 10^{-3} \text{ V}$$

It is noticeable that in this configuration the noise is exclusively dominated by the term e_{in}. In the second place is the contribution of R_2. But, for example, If R_2 = 1GΩ, 100 times bigger, the equivalent output noise in R_2 would be still lower (4.06×10^{-6}) than e_{in} (3.78×10^{-3}). Here noise is clearly from the OP-AMP and not from the external components, like photodiode or resistors.

In conclusion, noise is dominated by mainly e_{in} parameter. So, when noise is a critical in the design a very low noise OP-AMP must be selected. Noise will affect this criterion only when input is lower or in the range of this noise. Otherwise it does not matter what OP-AMP is used.

Clearly, to reduce the total output noise as possible, R_2 value, and I_d current (dark current) of the photodiode must be as low as possible.

R_2, and the photodiode, are external components that can be always calculated or choose for convenience.

In according with the discussed before in the circuit of the figure 13 the photo current pulse must be higher than the dark current to ensure to pass the threshold fixed by the output noise, and in this way to detect the pulse in the output. Otherwise, pulse will be buried into noise and not change will be noticed.

Dr. Fernando J. Moutinho

2.8 Input Capacity (C_i)

Input capacity (C_i) is always present at the input of the OP-AMP and is very important when gain is dominated by capacitance instead of external resistance.

Input capacity is difficult to eliminate due to stray capacitance in the wiring components. Good OPAMP must have always a very low input capacitance, typically it is in the range of 1-4 pF.

Input capacity plays an important role in noise gain. For example, in the negative feedback configuration noise can be amplified if the feedback capacitor (C_f) is not well calculated comparing to the input capacity.

Continuing with negative feedback, in the inverter and non-inverter configuration the noise gain is:

$$A_{vn} = \frac{C_{in}}{C_f} \qquad (13)$$

Figure 14 shows an equivalent noise gain circuit for: (a) inverter, and (b) non inverter configuration.

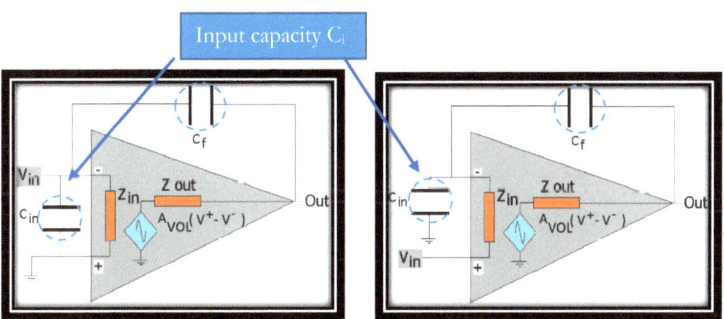

Figure 14. (a) noise gain in inverter model.   (b) Noise gain on non-inverter model.

Expression (13) shows that in both cases the feedback capacitance C_f must be higher than the input capacitance C_{in}, to keep noise gain as low as possible.

As a numerical example the figure 14 is showing a case of inverting amplifier with the following values:

R_F = 470.000 kΩ
R_1 = 470.000 kΩ

$C_{in} = 30$ pF
$C_f = 10$ pF

In this case the theoretical expected gain is $= |A_v| = \frac{R_f}{R_1} = 1$ and the cut frequency of the low pass filter is:

$$f_c = \frac{1}{2 * \pi * R_f * C_f} = 33.862 \; kHz$$

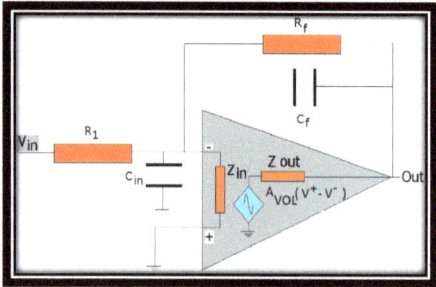

Figure 14. Inverting amplifier example.

Calculation has been made to plot the curve of the figure 15, 16 and 17.

Figure 15 shows the true gain curve including the noise gain and the theoretical one.

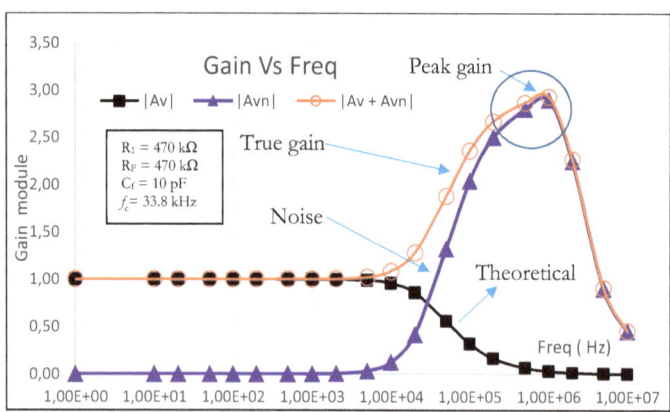

Figure 15. Plot of gain Vs frequency in the case of the figure 14.

The curve $|Av|$, in solid square, is representing the theoretical gain module response without any consideration of the input capacity (C_{in}). Curve $|Av_n|$, in solid triangle, is the noise gain module, and the $|Av+Av_n|$, in circle, is the real gain module in the circuit. It is observable that real gain is not rolling off at the expected cut frequency ($fc=33.8$ kHz), like the theoretical curve shows, instead the gain is increasing until it reaches the maximum value of ($\frac{C_{in}}{C_f} = 3$), then the total gain is attenuated as a function of the frequency, and by internal limitation in the OP-AMP (GBP).

This effect of increasing the gain, is known as peak gain an it is pointed in the figure 15.

Now, the figure 16 is showing the gain response as a function of the frequency obtained by changing the feedback capacitor ($C_f = 100$ pF) by a higher value.

Legends of the curves are the same.

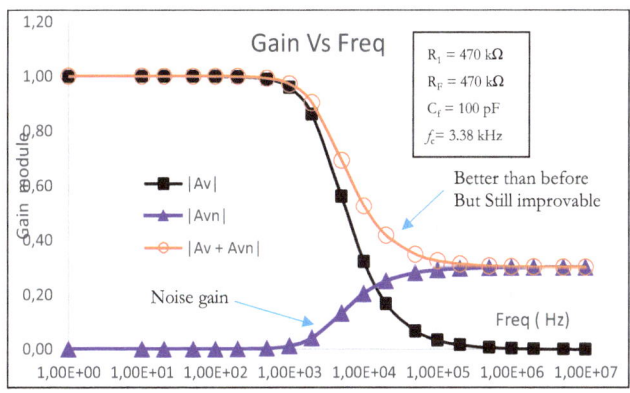

Figure 16. Plot of gain Vs frequency in the case of the figure 14. With $C_f = 100$ pF.

In the case of the figure 16 there is no longer the effect of the peak gain. Smoother curve of gain is obtained by changing just the C_f value. The price to pay is a reduction in the cut frequency of the circuit, now $f_c = 3.3$ kHz. But the overall response is closer to the expected by a good circuit.

Now, figure 17 is showing a circuit which re-calculated values so, the cut frequency(f_c) is maintained equal to 3.3 kHz by changing now R_1, R_f and C_F. Same gain calculations are made again, and the results curves are shown in the figure 17.

Looking at the figure 17, is clear to see that better result is obtained in this case. The noise curve is almost cero and the total gain $|Av + Avn|$ is very close to theoretical expected $|Av|$.

Again, the C_f value is much higher than C_{in}.

$$C_f > C_{in} \; ; to\ reduce\ peak\ gain\ and\ noise$$

In conclusion, C_f and C_{in} capacitors play an important role in the gain response and therefore, in the frequency stability of the OP-AMP circuit.

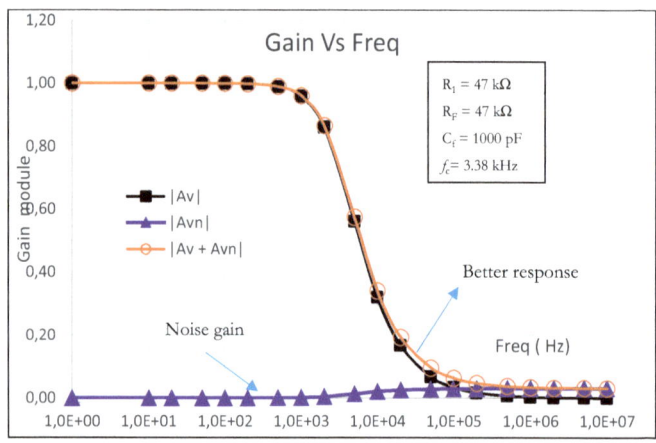

Figure 17. Plot of gain Vs frequency in the case of the figure 14. With C_f = 1000 pF.

In the example of figure 16 the reduction of the cut frequency was done intentionally, due to very high value of the input capacity (imposed) that makes necessary to choose a very high C_f value. Also, to change the value of the resistors. The idea is to demonstrate the effect of the input capacity in the gain behavior.

In normal cases C_{in} capacity is in the range of 1-4 pf, so making $C_f \geq 40$ pF, let's 100 pF, would be enough for the most of applications. But also the cut frequency can keep higher, by reducing the resistors values as for example 10 KΩ o lower. The same gain can be obtained with a combination of lower value resistors. Remember that higher resistance the higher the noise is. But also, the cut frequency decreases while resistance increases. So, it is always recommended to put resistance as low as possible.

Again, the idea is to make C_f capacitor always higher than C_{in} let`s say by a factor of 10. C_f must be always present in the circuit in parallel to R_f. Then using low values resistors as possible. This will avoid the peak gain effect appears in the gain response and keeping the intrinsically cut frequency limited by the OP-AMP parameter and not by external components.

Off course, C_f capacitor can be whatever value to reduce the cut frequency (low pass filter) by external component if desired, but always being higher than C_{in}.

2.9 CMRR (Common Mode Rejection Ratio)

In section one an introductory explanation was given about this parameter. The higher the CMRR, the better the OP-AMP is. The CMRR relation gives the idea on how much noise can be rejected compared with the signal. Typical CMRR value is in range between 90-120 dB, and it is defined as:

$$CMRR = 20 \log \left(\frac{A_{vd}}{A_{vc}}\right) \qquad (14)$$

As it was already mentioned before to keep noise gain at minimum it is necessary to use differential, instead single configuration, like is depicted in the figure 18.

To take advantages of the CMRR try to use differential configuration as possible. But nevertheless, the single configuration can be used in all the cases where signal to noise ratio (SNR) is very big, or in other words, signal is very strong, and noise is or can be neglected.

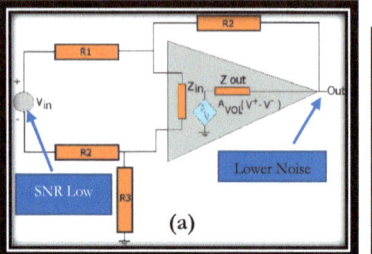
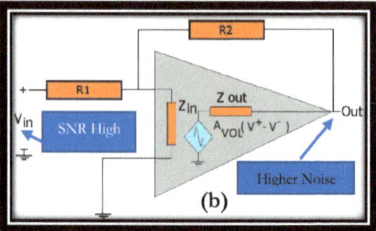

Figure 18. (a) Differential mode, (b) single mode.

Figure 19 is now representing the equivalent noise model for both: differential a single mode amplifier.

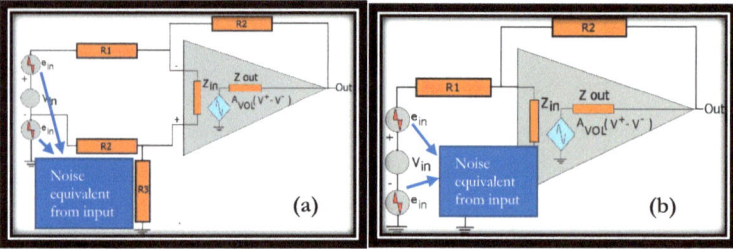

Figure 19. Noise model for: (a) Differential mode, (b) Single mode.

In the case (a) of the figure 19:

$$v_{out} = A_{VD}v_{in} + e_{in}A_{VC} \qquad (15)$$

Where:

$A_{VD} = the\ differential\ gain\ ; \gg 1$
$A_{VC} = the\ common\ gain;\ \ll 1$
$e_{in} = noise\ input$
$V_{in} = input\ signal$

In the case (b):

$$v_{out} = A_v(v_{in} + \Sigma e_{in}) \qquad (16)$$

Where: $\Sigma e_{in} = \sqrt{(e_{n0})^2 + (e_{n1})^2}\ ...$

Now comparing both equation (15) and (16) is easy to know that noise contribution is much lower in equation (15), due to the term A_{VC} is much lower than 1.

Dr. Fernando J. Moutinho

Thus, differential mode rejects better the noise, then is more suitable configuration in case where noise must be high rejected, very low SNR or in case of high gain amplifier.

It is proven that using the differential configuration we expand the advantages of the CMRR.

2.10 Input Bias Current.

In theory no current must flow between + and − OP-AMP ports.

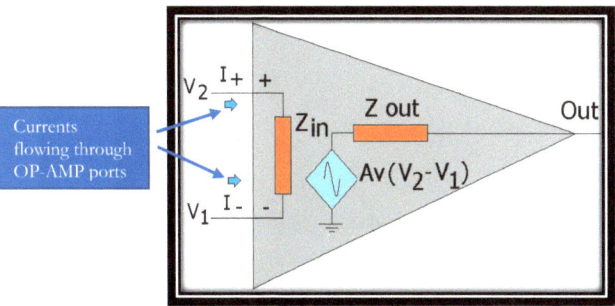

Figure 20. Input Bias current.

But it is not. Bias Current in the OP-AMP is also called I_{B+} and I_{B-} respectively. In practice, this current typically fall in the order of nanoamps to picoamps, but some special OP-AMPs can even get lower to femtoamps (10^{-15} A).

To see the importance of this current let's suppose the case in which we have an input bias current of 1 mA (intentionally). Look at the figure 21.

In the example of the figure 21 the I_{R2} current is:

$$I_{R2} = I_{R1} - I_B$$

Now, considering the present values: $V_{in} = 10$ V, $R_1 = 10k\Omega$, and $R_2 = 20$ kΩ

And let's also suppose that A_{VOL} = very high number (500.000 for example)

If $V^- \sim 0V$, can be neglected respect to V_{in}, then:

$$I_{R1} \sim \frac{V_{in}}{R_1} = 1\ mA$$

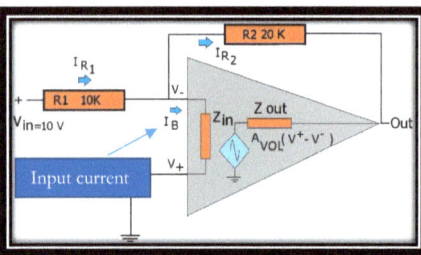

Figure 21. Considering the input Bias current.

Then: $I_{R2} = 0\ A$, what results:

$$V_{out} = -I_{R2} * R_2 = 0V$$

This simple calculation proves the fact that input bias current must be as low as possible, in all cases.

Now, assuming I_B = 10 nA for example, and recalculating in the same example:

$$I_{R2} = I_{R1} - I_B = 0.999\ mA$$

Then:

$$V_{out} = -I_{R2} * R_2 = -19.998\ V$$

The previous value is now matching with the expected value for estimated gain = -2

Normally, the user don't have to pay attention to this parameter since as stated before this current is typically in range of nA to pA. But nevertheless, in some cases where the input flow current (I_{R1}~ I_{in}) is very small or comparable to the input bias current then, it will be necessary to select a proper OP-AMP with an input bias current lower enough than the I_{in} current. This can be the cases of the electrometers, current meters (nanoamp), photocurrent detectors, as for example.

In any case to work properly the current flowing by the feedback resistor must be higher than input bias current.

OP-AMPs with an input bias current in order of femtoamps can be very expensive and sensitive to the noisy environments. So, they need to be very shielded from all noise sources.

2.11 Input Offset Current (I$_{OS}$)

The difference between the Input bias currents I$_{B+}$ and I$_{B-}$ is the input offset current:

$$I_{OC} = I_{B+} - I_{B-} \qquad (17)$$

In most cases the manufacture expresses the I$_{OS}$ in terms of minimum, typical and maximum values. This is because it is difficult to get a perfect matched input in all conditions. Let's say: temperature, voltage, application, etc, affect this parameter. In some cases, this difference can be tabulated on the OP-AMP datasheet as an absolute value, or as +/- values.

Some OP-AMPs have internal bias current compensation to improve this matching.

Figure 22 is representing an amplifier with external resistor used to improve the effects of input offset bias current.

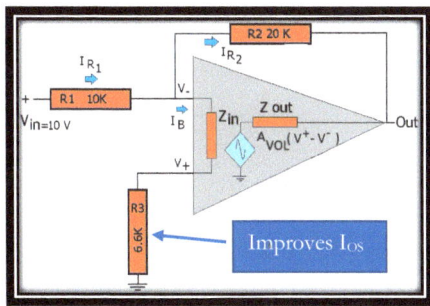

Figure 22. Canceling I$_{OS}$ with external resistor R$_3$.

In the circuit of the figure 22:

$$(V^+ - V^-)A_{VOL} = V_{out} \qquad (18)$$

And:

$$V^+ = -I_{B+}R_3 \qquad (19)$$

And:

$$\frac{(V_{in} - V^-)}{R_1} - I_{B-} = \frac{(V^- - V_{out})}{R_2} \qquad (20)$$

From equation (20):

$$V^- = \frac{R_1 V_{out}}{R_1 + R_2} + \frac{R_2 V_{in}}{R_1 + R_2} - \frac{(R_2 R_1) I_{B-}}{R_1 + R_2}$$

Then substituting in equation (18):

$$\left(-I_{B+} R_3 - \frac{R_1 V_{out}}{R_1 + R_2} - \frac{R_2 V_{in}}{R_1 + R_2} + \frac{(R_2 R_1) I_{B-}}{R_1 + R_2}\right) A_{VOL} = V_{out}$$

$$V_{out} = -\frac{R_2}{R_1} V_{in} + \frac{(R_2 + R_1) R_3}{R_1} I_{OS} \qquad (21)$$

Now considering R_3 is a parallel of R_1 and R_2:

$$R_3 = \frac{R_1 R_2}{R_1 + R_2}$$

$$V_{out} = -\frac{R_2}{R_1} V_{in} + R_2 I_{OS} \qquad (22)$$

Notice the convenience of R_3 being a parallel of R_1 and R_2, it makes de voltage dropped in V+ and V- due to the input bias current to be similar. The offset voltage due to this current is minimized to term $R_2 I_{OS}$ which is in any case results lower than obtained with $R_3 = 0\ \Omega$.

In the case of $R_3 = 0\ \Omega$.

$$V_{out} = -\frac{R_2}{R_1} V_{in} + R_2 I_{B-}$$

The $R_2 I_{B-}$ value is always greater than $R_2 I_{OS}$ because I_{B+} and I_{B-} are generally very similar.

This technique can perfectly apply in the case where the input bias currents are not well matched or not compensated. In the rest of the cases if the input bias current is compensated this criterion does not apply, so, R_3 can have any value. Designer must check this parameter in the manufacturer datasheet.

Dr. Fernando J. Moutinho

2.12 Input offset Voltage (V_{IO}, V_{OS})

Due to the difference between the input bias current (input offset current) and the finite input impedance, and others parameters in the OP-AMP design, the output is most of the time not exactly to 0 V, but very close, when it must be at 0 V. Normally, this offset in the output voltage is in the range of some millivolts or microvolts. Then, it is necessary to apply externally a certain amount of DC voltage to a one port V^- or V^+, in such way that the input difference compensates the output voltage to reach 0 V.

There are types of OP-AMP with auto compensated input offset voltage, so there is not needed to compensate externally by using variable resistor.

The external compensation is generally made by a series combination of a fix and variable resistor which are specifically for each OP-AMP manufacturer, allowing adjust the offset output voltage in the level to µV to mV.

Consideration must be taken with temperature dependence that affects this parameter as well as the input bias current. Drifting in the offset output voltage will be present if temperature changes along wide range.

The ideal input offset voltage is 0 V.

2.13 Output Current (I_O, OC)

Typically, OP-AMPs does not supply significate amount of current. Most of them provide a few tens of milliamps. So, may be, it can turn on a single led, but if it is needed to pull out more current to drive a higher load, like a relay for example, it needs to buffer the output current by a transistor, using one or two, depending if the output voltage swings a single or dual supply respectively.

Also, because output current is limited to a maximum value, a short circuit protection is indicated in this case (I_{SC}). The duration of the short circuit condition can be in some cases indefinitely in time. This is an internal protection in the OP-AMP that preserves the output to be burn by excessive heating. During the short circuit protection, the output can be folded back to a lower level voltage.

Is important to keep in mind that output current limit is conditioned to a pure resistive load. To drive inductive or capacitive loads is necessary to ensure that some protections are taken to avoid peak of current, overload, that can damage the output stage by transient. Also is recommended to check datasheet to look for this detail.

In the case of driving relays or any inductive loads is always recommended to use diode protection against the transient response.

2.14 Power Supply Rejection Ratio (PSRR)

The PSRR is defined as the ratio of the variation in supply voltage per variation in the output of the OP-AMP. It depends also of the feedback loop. There is no industrial standard definition on this term. But in theory, the bigger the PSRR the better is the rejection of the OP-AMP to the ripple or noise coming from the power supply.

$$PSRR\ (dB) = 20log_{10}(\frac{\Delta V_{supply}}{\Delta V_{out}}) \qquad (23)$$

Or:

$$PSRR\ (dB) = 20log_{10}(\frac{Ripple_{Input}}{Ripple_{Output}}) \qquad (24)$$

Generally, it is important to feed the OP-AMP with a good power supply, because noise or ripple can affect the output. The PSRR factor ensures a constant attenuation, but output will be affected according to the quality of the power supply.

3.0 Some practical circuits

The following section contents a technical guide for a good understanding and design technique of the most practical circuits that are used all the time.

The OP-AMP most common application is without doubt the amplifier.

In the context of a good design all is important: proper powering, decoupling filters, type of chosen OP-AMP, signal conditioning, noise treatment, filtering, distortion, and of course the amplification.

3.1 Powering the OP-AMP

The first thing to do is to decide if single or double power supply will be used. Most of the OP-AMPs can work with single or double power supply, but some of them do not well. So, it is necessary to consult the datasheet in any case before the design work.

If the cost is not a problem and reference to zero 0V is required, then dual supply is the option. But, if reference can be different from 0V, then single supply can be the chosen one.

Most of the time single supply is preferred because practical reasons: unique power supply instead of two, lower cost, and compact design.

Figure 23 is showing the pin configuration for most of the 8-pin single OP-AMP integrated circuit. However, it is recommended to verify the pin configuration prior to connect the OP-AMP.

Note that figure 23 is showing AC decoupling capacitors of typical value of 100 nf. This technique is highly recommended to reduce noise an inductance peaks coming from wiring and the power supply. With preference this capacitor must be made of Teflon, polypropylenc, polyester, or plastic type of dielectric to ensure very low equivalent series resistor (ESR) and equivalent series inductance (ESL). In this category are: silicon, mica, and ceramic ones.

Also, these capacitors must be placed physically as close as possible from the OP-AMP power supply pins. The ground path (capacitors) must have as big area as possible and be kept short in its trajectory to the source.

It is also possible to increase this capacitance (5-10 µF) by adding an electrolytic capacitor always in parallel with the reference 100 nF.

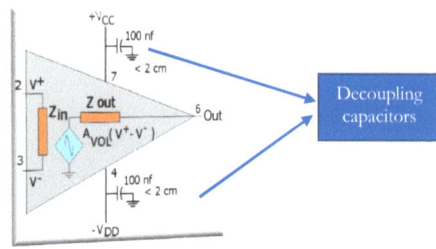

Figure 23. Powering the OP-AMP: dual power supply.

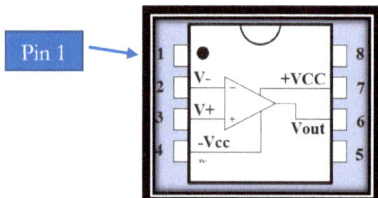

Figure 24. Typical 8 pin OP-AMP configuration pins.

Figure 25. Picture of 8-pin OP-AMP. In the picture the UA741 from Texas.

From this point forward every time an OP-AMP is used the configuration of the decoupling capacitors are implicit, so even when the circuit doesn't show it is supposed. In the case of single supply only one decoupling capacitor is needed, figure 26 shows an example. Remember, from here this capacitor will not be shown but it is implicit.

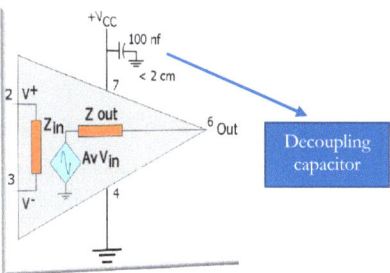

Figure 26. Powering OP-AMP with a single supply. Decoupling capacitor is shown.

Now let´s suppose single power supply is the choice now.

For positive power supply use the configuration as depicted in the figure 28.

Note that in the negative voltage the pin assigned to -V_{CC} is connected to ground.

When using a single supply, the negative swing is cut to ground level. Therefore, to amplify AC signals, is necessary to mount a DC level in the output of the OP-AMP. This DC level will be shifted by the AC input signal up and down, from de DC level to some voltage near to power supply, and from DC level to ground making the swing of the output. If symmetrical swing of the output is desired, then DC level must be equal to half of the power supply voltage.

$$Q_{DC} = \frac{V_{CC}}{2} \qquad (25)$$

Maximum symmetrical AC swing is:

$$V_{outmax} = 2Q_{DC} \qquad (26)$$

To avoid some clipping in the maximum output signal is necessary that Q_{DC} and gain are very stable with temperature and frequency.

3.2 Inverter amplifier

Figure 27 is showing now an a very common inverter amplifier with dual power supply to amplify AC signals.

Alternatively, figure 28 is showing same inverter amplifier but using single power supply to amplify AC signals.

Note that in figure 28 a DC level is necessary to amplify positive and negative input signal.

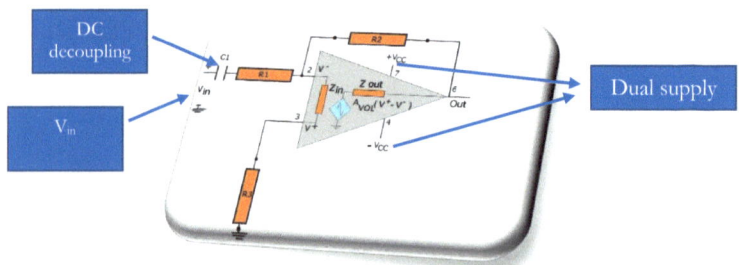

Figure 27. Dual supply AC amplifier. Power supply decoupling capacitors are implicit.

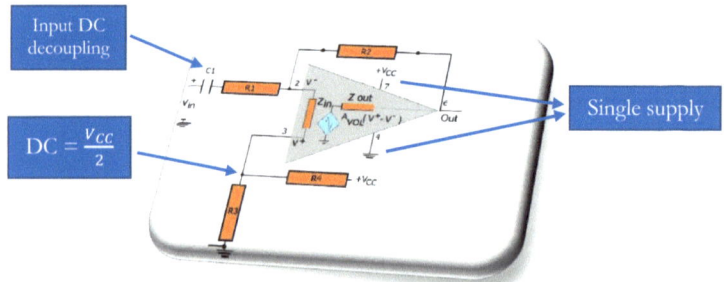

Figure 28. Single supply AC amplifier. Decoupling capacitor is implicit.

The amplifier in the figure 28 is decoupled in DC by C_1 capacitor.

As commented before R_3 is equal to parallel of $R_1//R_2$. (for a non-bias current compensated OP-AMP).

$$R_3 = R_1//R_2$$

Then:

$$V^+ = \frac{R_3}{(R_3+R_4)}V_{CC} = \frac{V_{CC}}{2} \quad (27)$$

$$R_4 = R_3$$

$$(V^+ - V^-)A_{VOL} = V_{OUT} \quad (28)$$

Where:

$$V^- = V^+ - \frac{V_{out}}{A_{VOL}}$$

Term: $\frac{V_{out}}{A_{VOL}} \sim 0$, then:

$$V^- \sim V^+ \quad (29)$$

Then:

$$\frac{(V_{in}-V^+)}{R_1} = \frac{(V^+ - V_{Out})}{R_2} \quad (30)$$

In **DC**:

$$\frac{(V^+ - V_{Out})}{R_2} = 0 \; ; \; V^+ = V_{Out} = \frac{V_{CC}}{2} \quad (31)$$

In DC mode the current in the feedback loop is 0 A because C_1 capacitor is decoupling the DC current in R_1.

Also note that in DC the output of the OP-AMP is clamped to DC level imposed by terminal V^+.

In **AC** (Note that input signal is fed through a decoupling capacitor)

Assuming $XC_1 \rightarrow 0$, in the work frequency region.

The equation (30) can be written like:

$$\frac{(v_{in})}{R_1} = \frac{(-v_{Out})}{R_2} \quad (31)$$

Where:

$$v_{Out} = -\frac{R_2}{R_1}v_{in} \quad (32)$$

Thus, the total superimposed (DC+ AC) response will be:

$$V_{Out} = -\frac{R_2}{R_1}v_{in} + \frac{V_{CC}}{2} \quad (33)$$

The equation (33) is valid for all type of input forms.

Figure 29 is showing now an improved version of the circuit in figure 28:

Note that capacitors C_2, C_3, and C_4 have been added to limit the bandwidth and to compensate the gain peak. Also, noise $\sqrt{Hz}$ will be reduced.

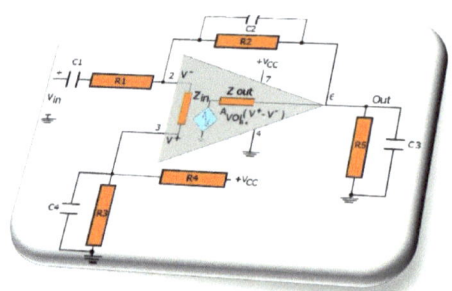

Figure 29. Improved version of figure 28 circuit.

The output cut frequency (f_{cL}) of the low-pass filter will be:

$$f_{cL} = \frac{1}{2\pi R_2 C_3} \text{ ; low-pass filter} \quad (34)$$

$$C_2 = \frac{1}{2\pi R_2 f_{cL}} \quad (35)$$

And the input cut frequency (f_{cH}) of the high-pass filter will be:

$$f_{cH} = \frac{1}{2\pi R_1 C_1} \text{ ; high-pass filter} \quad (36)$$

$$C_1 = \frac{1}{2\pi R_1 f_{cH}} \quad (37)$$

The filter bandwidth (BW) will be: $BW = f_{cL} - f_{cH}$ \quad (38)

C_4 is a DC capacitor for stabilizing the DC level in OP-Amp port and can be calculated as low pass filter with a cut frequency very low, as 50-60 Hz for example.

C_3 and R_5 play more important role and should be recommended values for optimal SR and smoothing the signal. Typical values ere hare: 100 pF and 2 kΩ respectively.

Note: the output of amplifier can be DC decoupled by using a series capacitor. Then DC can be removed from output using a series decoupling capacitor.

The transfer function graph of this amplifier is shown in the figure 30, and the equation (39) shows the mathematical expression from what the graph of the figure 30 is obtained.

$$|H(w)| = \frac{R_2}{wC_2(\sqrt{(R_2 R_1 - \frac{1}{w^2 + C_1 C_2})^2 + (\frac{R_2}{wC_1} + \frac{R_1}{wC_2})^2})} \tag{39}$$

Where: $w = 2 * \pi * f$

Equation (39) shows only absolute values but keep in mind that gain is negative in this circuit. So, output signal will be 180º phase shifted in relation to the input signal.

For numerical results some values have been given:

R_1= 1kΩ, R_2= 20Ω, C_1= 7.9µF, C_2=398 pF.

As expected, the output of the amplifier behaves as a band-pass filter. The maximum gain occurs at the square root of f_{cL} and f_{cH}:

$$f = \sqrt{f_{CL} * f_{cH}} \tag{40}$$

The maximum gain of the amplifier is 20 and the bandwidth is about 20 KHz.

$$BW = f_{CL} - f_{CH} = 20.000\ Hz - 20\ Hz \sim 20.000\ Hz$$

The graph of figure 30 shows the gain in dimensionless units as a function of the frequency in a semilogarithmic scale. The figure 31 is the same curve of figure 30 but using a normalized gain scale.

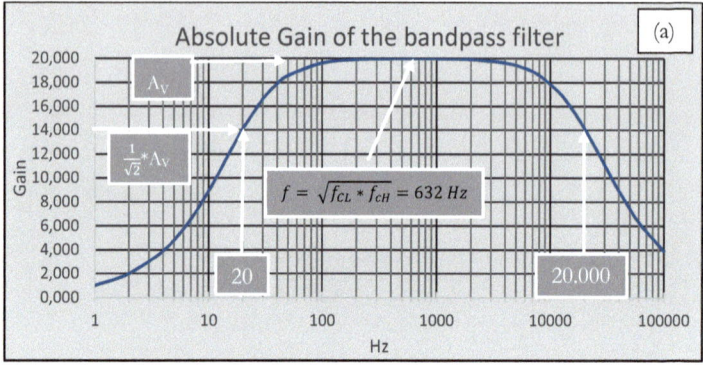

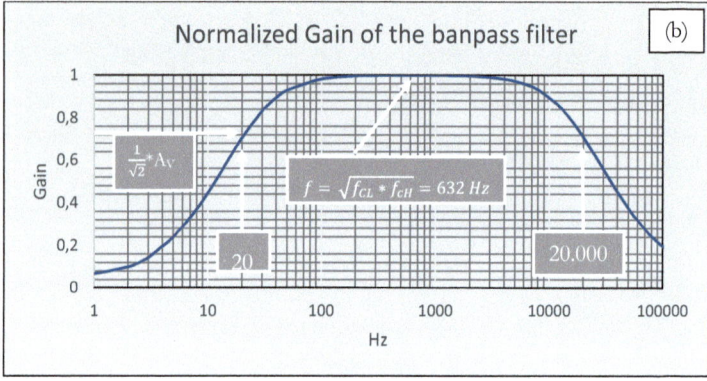

Figure 30. (a) Absolute Gain response $|H(w)|$ for the amplifier of figure 30. (b) Normalized Gain plot of figure 30(a).

Graphical response shows a band pass filter with a wide region of passing frequencies.

In the regions below and above respective cut frequencies signal is attenuated as a function of the frequency. The rate of attenuation is a constant. The response over the band pass can be taken in approximation as a flat, thus a constant gain.

As it is depicted in figure 31 the roll off frequency gives attenuation of -20 dB/decade. According to normal first order RC filter.

Figure 31 is showing the gain in dB units as a function of the frequency.

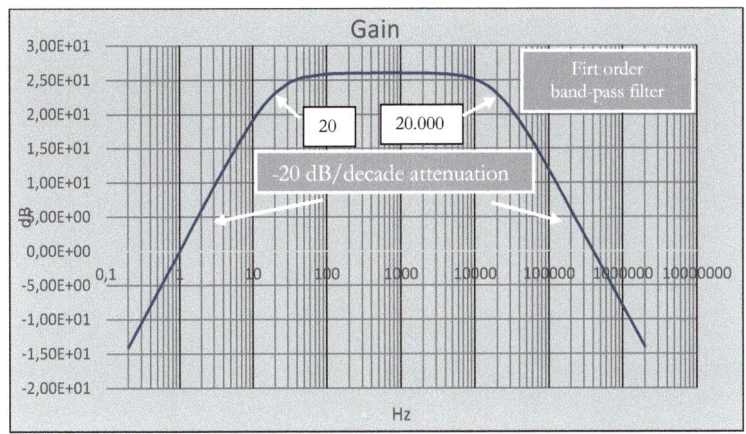

Figure 31. Decibels equivalent amplitude response from figure 30.

As it is pointed in figure 30 the cut frequencies are determined where the gain is reduced to: $\frac{1}{\sqrt{2}} = 0.707 = 70.7\,\%$, from its maximum.

Figure 32 is now showing the gain output plot from two input signals:

- $F_1(t) = 0.25 * \mathrm{Sin}(2 * \mathrm{pi}() * f * t);\ f_1 = 1$ kHz
- $F_2(t) = 0.25 * \mathrm{Sin}(2 * \mathrm{pi}() * f * t);\ f_2 = 25$ kHz

inputs

Where: f = is the frequency in kHz of the input signal, and t is time in milliseconds and $\mathrm{pi}() = \pi$

$$w = 2 * \pi * f \qquad (41)$$

$$wt = 2 * \pi * f * t \qquad (42)$$

It is also supposed that both signals are in phase, and with identical amplitudes. The only difference is their frequency.

What figure 32 is showing is the output of the two referred input signals.

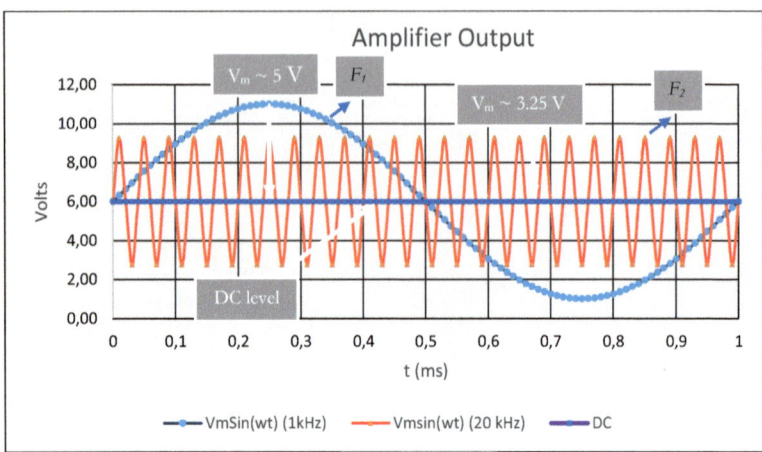

Figure 32. Output response from two input signals. Case: figure 28 amplifier.

Equation (39) can be tabulated, in excel for example, then evaluated as a function of the frequency to know very well the gain value for each frequency value. However, from the plot of the figure 30 these values can be also extrapolated.

From the figure 30 is easy to see the gain for 1 kHz input signal, it is almost ~20, because it is within the center of the band-pass filter. The gain at this frequency is very close to maximum. The maximum gain is obtained when frequency is at 632 Hz, as indicated by the equation (40). The exact gain value must be obtained from evaluating the expression (39). But for this purpose, will be enough using the curve of figure 30 for approximation. Then the gain of the amplifier here is roughly the maximum, then:

$$V_{Out} = A_V * F_1(t) \sim 5\sin(wt) \, ; A_V \sim 20$$

And using the same approximation from curve of figure 30, in the case of 25 kHz, $A_V \sim 13$, then:

$$V_{Out} = A_V * F_2(t) \sim 3.25\sin(wt) \, ; A_V \sim 13$$

This example servers as demonstration of frequency dependency of the gain. But also, to show the use of the filter function in amplifying signal.

As it was mentioned before the use of the bandpass filter reduce also the $\sqrt{Hz}$ noise.

In the case of the 25 kHz input signal which clearly is out of band-pass filter, the output is attenuated with a ratio of -20 dB/decade.

Frequency is attenuated with -20 dB per decade below and over the respective cutting frequencies.

Every -20 dB is equivalent to $\frac{1}{10}$ attenuation. Therefore, in 2 decades the total attenuation is -40 dB, equivalent to $\frac{1}{100}$ attenuation.

For example, for 2 MHz input signal the total expected attenuation in the output will be -40 dB, it means the gain at this frequency is:

$$A_{V|2MHz} = A_V * 0.01 = 0.2$$

If: $F_3(t) = 0.25*\mathrm{Sin}(2*\mathrm{pi}()\ *f*t)$; $f_3 = 2$ MHz

The output voltage will be:

$$v_{Out} = A_{V|2MHz} * F_3(t) = 0.05 \sin(2 * pi() * f * t)$$

In the figure 31 this output corresponds to:

$$Gain_{|2MHz} = 20 \log(A_{V|2MHz}) \sim -14\ dB$$

The above result can be checked also form the figure 31.

Now to complete the design is necessary to have a proper GBP parameter. As it was discussed before, gain can be modified for this parameter.

The upper work frequency is the key here. For 20 kHz (higher frequency), that by coincidence is the bandwidth here, and a gain of 20, its product results in 400 kHz. But, with this GBP the circuit will suffer a second order attenuation at 20 kHz, because OP-Amp limit is operating at the same frequency. Then it necessary widen the previous value for at least a factor of 10 to ensure no attenuation from OP-AMP within the work region.

$$GBP \geq 10 * f_{CL} * A_V \qquad (43)$$

Then: $GBP \geq 10 * f_{CL} * A_V = 10 * 20\ kHz * 20 = 4\ MHz$

Here f_{CL} is the higher frequency, belongs to the low pass filter.

Dr. Fernando J. Moutinho

Then a 4 MHz bandwidth OP-AMP must be chosen.

Others parameters are likely will not affect the overall design, so it can be fitted by most of the OP-AMP in the market.

Summary:

In the design of a good amplifier a bandpass filter must be always considered. This will result in less noisy and more reliable design. Also, using no very high resistors or gain. GBP parameters must be also checked as well others already mentioned parameters.

Next part we present a table with all formulas for this circuit. The purpose of this is to reduce time consuming in the design by tabulating the most important parameters, additionally to ensure a good design.

Project: Inverter amplifier with filter

[Circuit diagram: Single supply inverter amplifier with filter, showing V_{in}, C_1, R_1, R_2, C_2, op-amp, R_3, R_4, R_5, C_3, $V_{CC}/2$ supply.]

$R_5 = 2\ \text{k}\Omega;\ C_3 = 100\ \text{pF}$

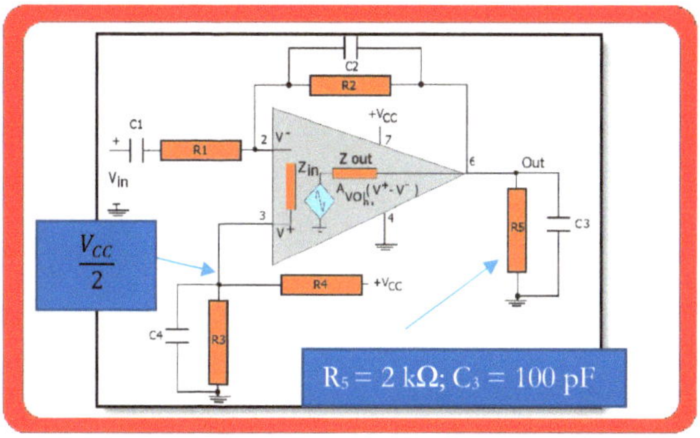

Parameter 1: ☐

$$v_{Out} = -\frac{R_2}{R_1} v_{in} + \frac{V_{CC}}{2} \qquad (33) \quad R_4 = R_3$$

Parameter 2: ☐

$$f_{cL} = \frac{1}{2\pi R_2 C_3}\ ;\ \text{low pass filter} \qquad (34)$$
$$C_2 = \frac{1}{2\pi R_2 f_{cL}} \qquad (35)$$

Parameter 3: ☐

$$f_{cH} = \frac{1}{2\pi R_1 C_1}\ ;\ \text{high pass filter} \qquad (36)$$
$$C_1 = \frac{1}{2\pi R_1 f_{cH}} \qquad (37)$$

Parameter 4: ☐

$$BW = f_{CL} - f_{CH} \qquad (38)$$
$$f = \sqrt{f_{CL} * f_{cH}}\ ;\ center\ frequency \quad (40)$$

Parameter 5: ☐

$$GBP\ (OP-AMP) \geq (10 * f_{CL} * A_V)$$

Comments:

Single Supply inverter with band-pass filter included.   **Single Supply**

Table 1. Smart Summary of inverter amplifier design

3.3 Non-Inverter amplifier

Figure 33 is showing a model of a non-inverter amplifier.

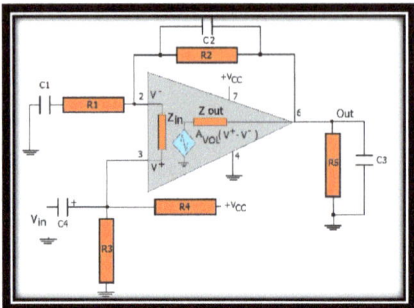

Figure 33. Non-Inverted amplifier with band-pass filter.

Similar than inverter amplifier the output is shifted to DC level equal to:

$$\frac{R_3}{R_3 + R_4} V_{CC} = \frac{V_{CC}}{2} \mid R_4 = R_3$$

For calculating the gain, we have next equations:

$$(V^+ - V^-) * A_{VOL} = V_{Out} \qquad (44)$$

$$V^+ = V_{in}$$

Assuming X_{C1} and X_{C2} impedances are neglected in this part of the calculation. So, in the work region $X_{C1} \rightarrow 0$, and $X_{C2} \rightarrow \infty$ for the high pass and low pass respectively.

Then:

$$\frac{-V^-}{R_1} = \frac{(V^- - V_{Out})}{R_2} \qquad (45)$$

Where:

$$V^- = \frac{R_1 V_{Out}}{(R_1 + R_2)}$$

Then substituting in (44):

$$A_V = \left(1 + \frac{R_2}{R_1}\right) \tag{46}$$

The overall output will be (AC + DC):

$$v_{Out} = \left(1 + \frac{R_2}{R_1}\right) v_{in} + \frac{V_{CC}}{2} \tag{47}$$

Note that gain has a minimum value of 1. In the case of the inverter the gain can reach even the value close to 0. Also, the phase sign here is positive, so the output and input are in the same phase.

The frequency response is very similar than the case before except that transfer function has a constant value added. This means that when frequency tends to infinite the gain will be asymptotic to 1. The effect of residual gain decreases the effective attenuation in dB of the filter. Figures 34 and 35 show this effect.

This effect is also more notorious when the gain is relatively close or comparable to unity gain.

As for example, for a gain: $R_2/R_1 = 100$ the factor 1 has poor or little effect within the first decade. But still the minimum gain will be 1, so it will affect the next decades until there will be no more attenuation. In others words, the final attenuation is 0 dB.

In the figure 34 the maximum gain is 21. The cut frequency for the low-pass and high-pass filters are calculated in the same way as the case before. They are defined identically. Also, the frequency for the maximum gain will be the square root of the product of both: low and high frequency.

Others parameters remains in the same way as in the inverter case.

Figure 34 and 35 have been plotted assuming same values as in the case of inverter. Particularly, in this case, the C_4 capacitor should be calculated in the way that:

$$X_{C4} \ll X_{C1} \; ; \; C_4 \gg C_1 \tag{48}$$

For numerical results some values have been given:

R_1= 1kΩ, R_2= 20Ω, C_1= 7.9μF, C_2=398 pF.

The transfer function of the gain is:

$$1+|H(w)| = 1 + \frac{R_2}{wC_2(\sqrt{(R_2R_1 - \frac{1}{w^2+C_1C_2})^2 + (\frac{R_2}{wC_1} + \frac{R_1}{wC_2})^2})} \text{; from Eq(39)}$$

The 0.707 value of $|H(w)| = 20*0.707 = 14.14$; then the cutting frequencies occurs at the gain = 15.14. Look at the figure 34.

Also look at the figure 35 the change in the attenuation rate from -20 dB/dec to 0 dB.

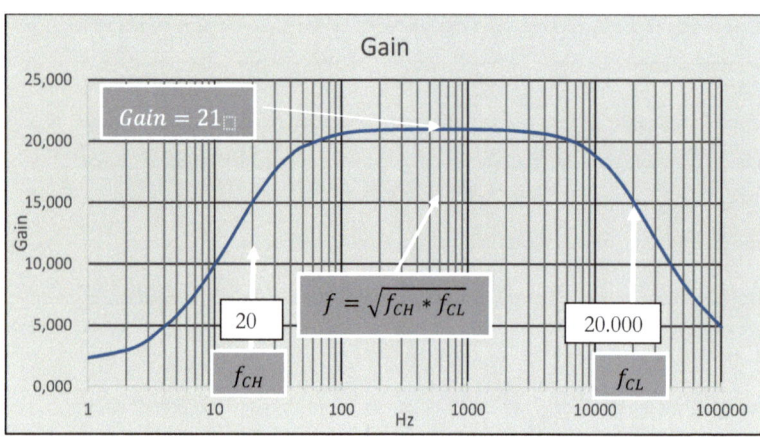

Figure 34. Gain response for the Non-Inverter amplifier.

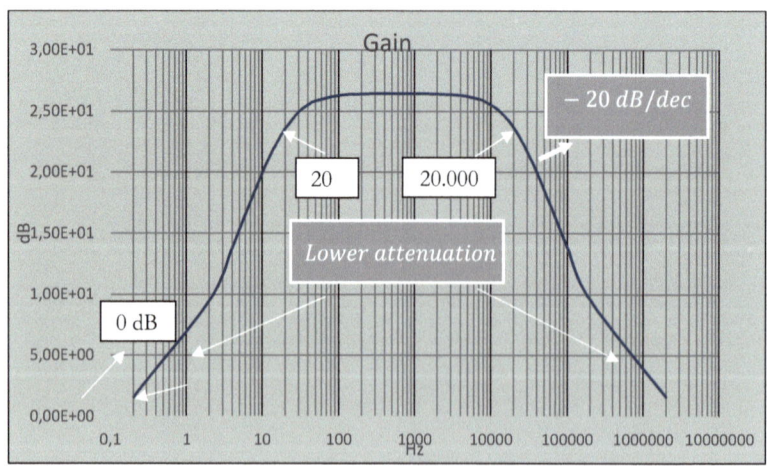

Figure 35. Gain response in dB of the Non-Inverter amplifier.

Dr. Fernando J. Moutinho

Project: Nom-Inverter amplifier

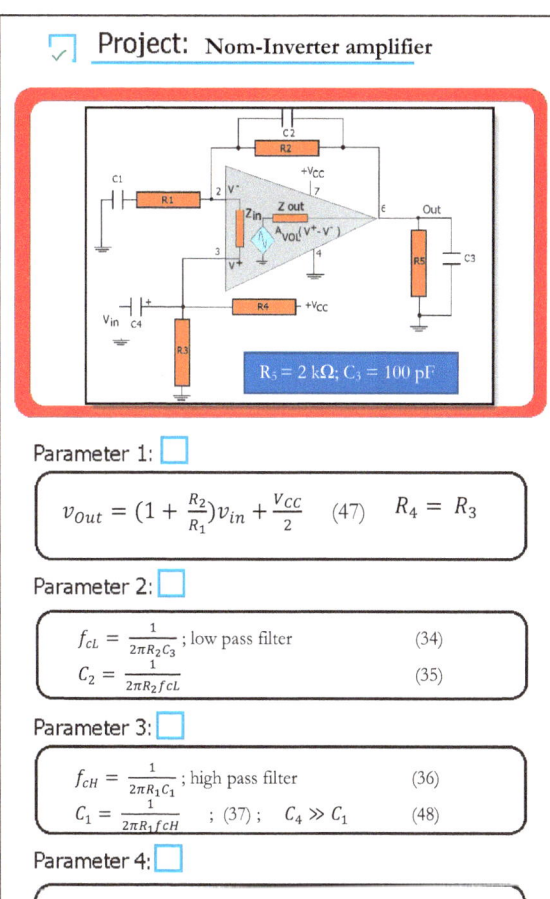

$R_5 = 2\ \text{k}\Omega;\ C_3 = 100\ \text{pF}$

Parameter 1: ☐

$$v_{Out} = \left(1 + \frac{R_2}{R_1}\right)v_{in} + \frac{V_{CC}}{2} \quad (47) \quad R_4 = R_3$$

Parameter 2: ☐

$$f_{cL} = \frac{1}{2\pi R_2 C_3}\ ;\ \text{low pass filter} \quad (34)$$
$$C_2 = \frac{1}{2\pi R_2 f_{cL}} \quad (35)$$

Parameter 3: ☐

$$f_{cH} = \frac{1}{2\pi R_1 C_1}\ ;\ \text{high pass filter} \quad (36)$$
$$C_1 = \frac{1}{2\pi R_1 f_{cH}}\ ;\ (37)\ ;\quad C_4 \gg C_1 \quad (48)$$

Parameter 4: ☐

$$BW = f_{CL} - f_{CH} \quad (38)$$
$$f = \sqrt{f_{CL} * f_{CH}} \quad (40)$$

Parameter 5: ☐

$$GBP \geq (10 * f_{CL} * A_V)$$

Comments:

Final gain is 1. Final attenuation is 0 dB.

Single supply

Table 2. Smart Summary of Non-inverter amplifier

3.4 Unity Gain (Buffer)

The unity gain amplifier is also known as buffer. As the name suggest the gain in this case is fixed to 1. Because the gain is 1, it takes advantage of maximum GBP. The bandwidth is highest, giving the maximum frequency range. But this type of circuit is used only in the cases of buffering or coupling impedance from one stage to another in the chain of amplifying. Figure 36 is showing the unit gain configuration.

To ensure the maximum bandwidth is advisable to use optimal C_3R_5 constant in the circuit, to improve SR parameter. Also, to use R_4, R_3 constant as low as possible to maintain match with V^- input port. Normally, the impedance of the input is this case is relatively high, so set first R_3 value, and then fix R_4 value. This value can be for example: 50 or 100 ohms for $R_4 = R_3$. The parallel of $R_3//R_4$ can be calculated to maximum power transfer from the input source resistance (R_S).

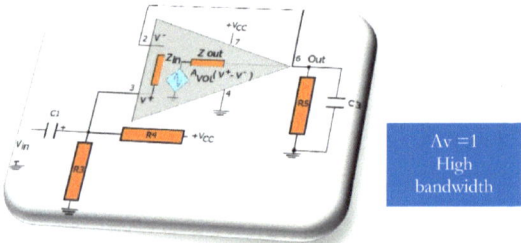

Figure 36. Unity gain amplifier (Buffer).

Note that there is a wire that shorts the output to V^- input of the OP-AMP.

$$(V^+ - V^-)A_{VOL} = v_{Out}$$

$$V^- = v_{Out}, and\ V^+ = v_{in}$$

Then:

$$v_{in} = V^+ = V^- = v_{Out} \qquad (49)$$

The upper limit frequency is this case is the GBP value.

The table 3 is showing a smart summary of this configuration.

Electronics: OP-AMP tech notes...! | 3.4 Unity Gain (Buffer)

☑ **Project:** Unity Gain Amplifier

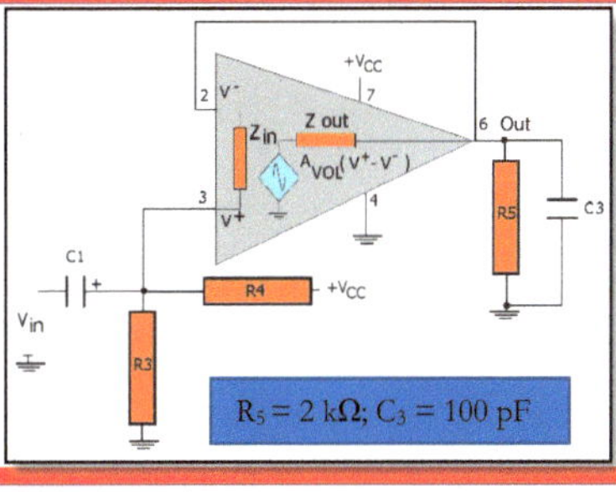

$R_5 = 2\ k\Omega;\ C_3 = 100\ pF$

Parameter 1: ☐

$$A_V = 1$$

Parameter 2: ☐

$$f_{max} = GBP$$

Parameter 3: ☐

$$R_3 = Rs\ \text{(from input source)}$$

Parameter 4: ☐

$$C_3 * R_5 = minimum\ to\ imprive\ SR$$

Parameter 5: ☐

$$V_{in} = V^+ = V^- = V_{out} \qquad (49)$$

Comments:

Single Supply

Table 3. Smart Summary of Unity gain amplifier

3.5 Inverter Sum Amplifier

The sum amplifier is one of most useful circuit, it can sum two or more signals. The sum can be in DC or AC regimen. Let´s study the case for two (2) signal in AC regimen. The process can be repeated for *n* input signals in DC or AC.

Figure 37 is showing the sum amplifier.

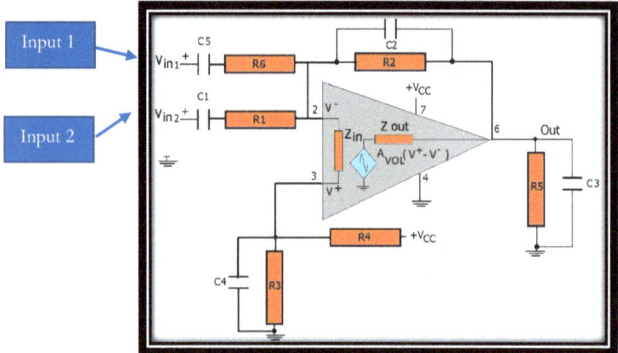

Figure 37. Sum inverter amplifier.

Notice that the C_1, C_5 capacitors are decoupling DC signals. The output (DC)of the amplifier will be equal to voltage applied in V^+ pin. The AC signal will only shift the DC level up or down according to the input swing.

In **DC**:

Using same procedure made in section 3.2:

$$V_{Out} = \frac{R_3}{(R_3 + R_4)} V_{CC}$$

In **AC**:

Applying superposition method and using again the calculation procedure established in section 3.2:

For the v_{in1}:

$$v_{Out} = -\frac{R_2}{R_1} v_{in1} \; ; \; v_{in2} = 0 \qquad (50)$$

For the v_{in2}:

$$v_{Out} = -\frac{R_2}{R_6} v_{in2} \; ; \; v_{in1} = 0 \qquad (51)$$

Then:

$$v_{Out} = -(\frac{R_2}{R_1} v_{in1} + \frac{R_2}{R_6} v_{in2}) \qquad (52)$$

Now adding all AC and DC components together:

$$v_{Out} = -\left(\frac{R_2}{R_1} v_{in1} + \frac{R_2}{R_6} v_{in2}\right) + \frac{R_3}{(R_3+R_4)} V_{CC} \qquad (53)$$

If $R_3 = R_4$ then:

$$v_{Out} = -\left(\frac{R_2}{R_1} v_{in1} + \frac{R_2}{R_6} v_{in2}\right) + \frac{V_{CC}}{2} \qquad (54)$$

Additionally, if: $R_1 = R_6$:

$$v_{Out} = -A_V(v_{in1} + v_{in2}) + \frac{V_{CC}}{2} \qquad (55)$$

Where: $A_V = -\frac{R_2}{R_1}$

Note that negative sign mean that AC output signal is inverted (180° phase shift) respect to the input.

Also, is noticeable that a band-pass filter is made with C_2, C_1, and C_5.

Low-pass filter frequency is:

$$f_{cL} = \frac{1}{2 * \pi * R_2 * C_2}$$

High-pass filter is:

$$f_{cH} = \frac{1}{2 * \pi * R_1 * C_1}$$

$$C_1 = C_5$$

Important: the sum of the signals will be performed only in amplitude. The output amplitude will be the input times by the Av factor. But to accomplish this, both signals must have the same response in the band-pass filter. That means that both signals are affected by the transfer function of the filter. So, if one the signals is attenuated different from the other (different frequencies) even when both have the same amplitude at the input, the result will be a two signal with a different amplitude each one.

Therefore, the corrected equation (56) is:

$$v_{Out} = -A_V(|H(w)|\ v_{in1} + |H(w)|\ v_{in2}) + \frac{V_{CC}}{2} \tag{56}$$

Where: $|H(w)|$ is the normalized transfer function module of the band-pass filter as a function of the frequency.

The module $|H(w)|$ is expressed in the section 3.2, also a band-pass filter.

The bandwidth is:

$$BW = f_{cL} - f_{cH}$$

When summing two signals both of then should be in the pass region of the filter, where the gain is steady and a phase shift is close to 0° otherwise, it is necessary to consider not only change in amplitudes but also in phase of the output response of the amplifier.

The concerned problem arises from the fact that f_1 and f_2 signals may have different frequencies. No problem in the case of same frequency.

Next is the table 4 with the smart summary for this case.

Project: Inverter Sum Amplifier

[Circuit diagram of inverter sum amplifier with components C_5, R_6, C_2, R_2, C_1, R_1, R_5, C_3, R_3, C_4, R_4, op-amp with $+V_{CC}$ supply, inputs V_{in1}, V_{in2}, and output Out. $R_5 = 2\ k\Omega$; $C_3 = 100\ pF$]

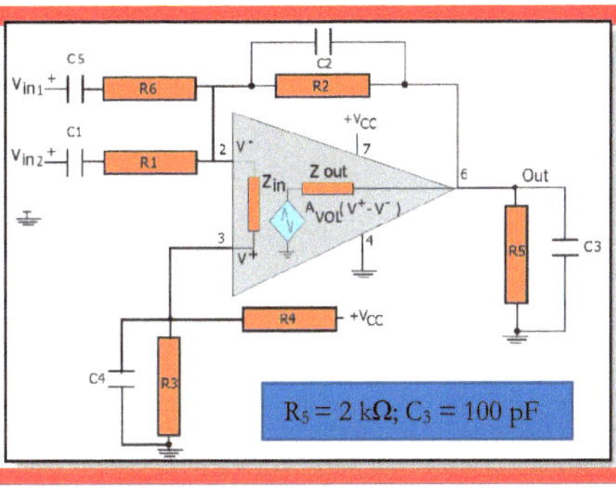

Parameter 1: ☐

$$v_{Out} = -A_V(v_{in1} + v_{in2}) + \frac{V_{CC}}{2} \quad (55)$$

Parameter 2: ☐

$$v_{Out} = -A_V(|H(w)|\ v_{in1} + |H(w)|\ v_{in2}) + \frac{V_{CC}}{2} \quad (56)$$

* Considering transfer function

Parameter 3: ☐

$$f_{cL} = \frac{1}{2*\pi*R_2*C_2}\ ;\ \text{low pass}$$

Parameter 4: ☐

$$f_{cL} = \frac{1}{2*\pi*R_2*C_1};\ \text{high pass} \qquad C_1 = C_5$$

Parameter 5: ☐

$$A_V = -\frac{R_2}{R_1}\ ;\ R_1 = R_6$$

Comments:
Where: $|H(w)|$ *is the normalized transfer function module of the band-pass filter.*

Single Supply

Table 4. Smart Summary of inverter sum

3.6 Differential Amplifier

Figure 38 is showing the differential amplifier circuit. This circuit is adapted to work with a single supply. But also works with dual power supply with minor changes. In fact, most t of the time is easier to work with dual power supply, since does not need any shifted DC level.

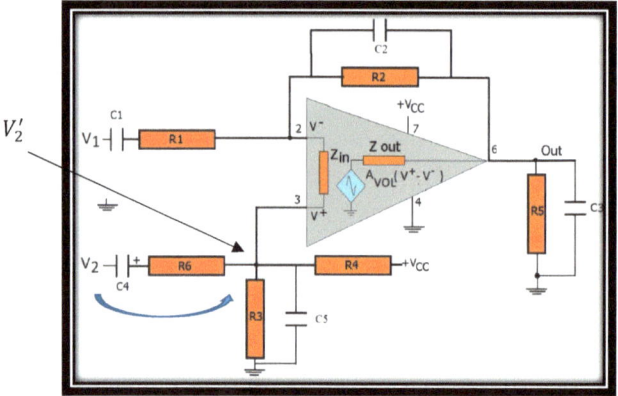

Figure 38. Differential amplifier.

In the circuit of the figure 38:

$$(V^+ - V^-)A_{VOL} = v_{Out} \qquad (57)$$

$$\frac{(V_1 - V^-)}{R_1} = \frac{(V^- - v_{Out})}{R_2} \qquad (58)$$

$$V^+ = v_2' \qquad (59)$$

From equation (58:)

$$V^- = \frac{R_1 v_{Out} + R_2 v_1}{(R_2 + R_1)}$$

Substituting in (57):

$$\left(v_2' - \frac{R_1 v_{Out} + R_2 v_1}{(R_2 + R_1)}\right) A_{VOL} = v_{Out}$$

Assuming: $\frac{v_{Out}}{A_{VOL}} \to 0$ then:

$$v_{Out} = \frac{(R_2+R_1)v_2' - R_2 v_1}{R_1} \qquad (60)$$

Now if: $R_1 = R_2$ and $R_6 = R_3 // R_4$:

$$v_2' = \frac{R_3 // R_4}{R_6 + R_3 // R_4} v_2 = \frac{1}{2} v_2$$

Replacing in (60):

$$v_{Out} = (v_2 - v_1) \quad (AC) \qquad (61)$$

Remember that a DC Level is imposed in V⁺ port so, equal that in previous cases the output is fixing in the same DC level. If $R_3 = R_4$ then the complete output is:

$$v_{Out} = (v_2 - v_1) + \frac{V_{cc}}{2} \quad (AC+DC) \qquad (61a)$$

Differentiator works in AC mode.

Note that in AC mode the impedance in v_2' is the parallel of R_3 and R_4. Therefore, the impedance seen by the source v_2 is:

$$Z_{inV2} = R_6 + R_3 // R_4 = 2R_6 \qquad (62)$$

Also, it is assumed for now that $X_{C5} \to \infty$ in work region frequencies.

Due to the parallel, if precision is required, resistor must in be order or 1% tolerance.

The high-pass filter cutting frequencies for V_1 and V_2 are respectively:

$$f_{C1} = \frac{1}{2\pi R_1 C_1}$$

And:

$$f_{C2} = \frac{1}{2\pi (2R_6) C_4}$$

And:

$$f_{C1} = f_{C2}$$

For the low pass filter cut frequency:

$$f_{C3} = \frac{1}{2\pi R_2 C_2}$$

The C_5 capacitor is required for compensating the gain in frequency. As it is shown in the equation (60), R_1 is one side of the transfer function. At high frequency the term of v_1 tends to be attenuated by the feedback impedance of R_2 and C_2. The same must be done with v_2 making a low pass filter with $(R_3//R_4)$ and C_5, so v_2 input must be attenuated in same way as v_1.

Then the low pass cut frequency is same as f_{C3}:

$$f_{C4} = \frac{1}{2\pi (R_3//R_4) C_5} = f_{C3}$$

So as long of signal V_1 is reduced the signal V_2 is also reduced.

Next is the table 5 with the smart summary for this case.

Electronics: OP-AMP tech notes...! | *3.6 Differential Amplifier*

✓ **Project:** Differential Amplifier

[Circuit diagram: Differential amplifier with f_{c1}, f_{c2}, f_{c3}, f_{c4} labeled at different nodes; components V_1, C_1, R_1, V_2, C_4, R_6, R_3, C_5, R_2, C_2, R_5, C_3, R_4; $R_5 = 2\ k\Omega$; $C_3 = 100\ pF$]

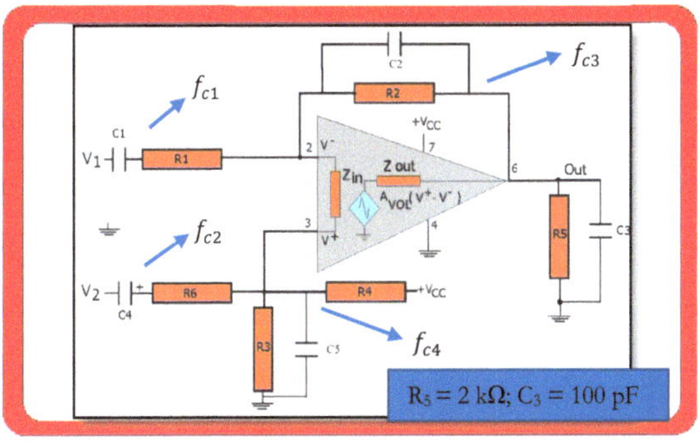

Parameter 1: ☐

$$v_{Out} = (v_2 - v_1) \quad R_1 = R_2 \text{ and } R_6 = R_3//R_4$$
(61) (AC)

Parameter 2: ☐

$$v'_2 = \frac{R_3//R_4}{R_6 + R_3//R_4} v_2 = \frac{1}{2} v_2$$

Parameter 3: ☐

$$Z_{inv2} = R_6 + R_3//R_4 = 2R_6 \quad (62)$$

Parameter 4: ☐

$$f_{C1} = \frac{1}{2\pi R_1 C_1}\ ;\ f_{C2} = \frac{1}{2\pi(2R_6)C_1}\ ;\ f_{C1} = f_{C2}$$

Parameter 5: ☐

$$f_{C3} = \frac{1}{2\pi R_2 C_2}\ ;\ f_{C4} = \frac{1}{2\pi(R_3//R_4)C_5}\ ; f_{C3} = f_{C4}$$

Comments:

The complete AC+DC signal is:

$v_{Out} = (v_2 - v_1) + DC$ *Single supply*

Table 5. Smart Summary of inverter sum

3.7 Inverter Integrator Amplifier

Figure 39 shows the basic integrator configuration in a single power supply configuration. Again, because a single supply is being used a DC level is required to drive the AC output.

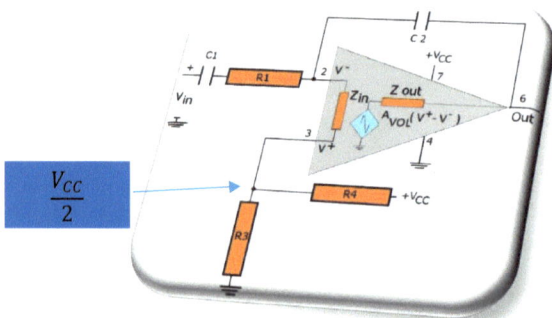

Figure 39. Integrator amplifier.

In the circuit of the figure 39:

Assuming: $V^+ = \frac{V_{CC}}{2}$; $R_3 = R_4$

$V^- = V^+ = \frac{V_{CC}}{2} = v_{Out}$ with $v_{in} = 0$

$$V_{C2} = (V^- - v_{Out}) = \frac{1}{C2}\int_0^t i_{C2}\, dt$$

$$v_{Out} = V^- - \frac{1}{C2}\int_0^t i_{C2}\, dt$$

$$i_{C2} = i_{R_1} = \frac{v_{in}}{R_1}$$

Thus substituting:

$$v_{Out} = \frac{V_{CC}}{2} - \frac{1}{R_1 C_2}\int_0^t v_{in}\, dt + V'_{C2} \qquad (63)$$

Notice that the integrator is also an inverter.

Where V'_{C2} = initial C_2 voltage state

For example, if $v_{in}(t)$ is a periodic square signal:

$$v_{in} = +V_m \Big|_0^{\frac{T}{2}} \qquad v_{in} = -V_m \Big|_{\frac{T}{2}}^{T} \qquad \text{and the Duty Cycle} = 50\%$$

And $|V_m| = \frac{V_{CC}}{2}$; square signal amplitude

In the first positive semicycle, integrating along: $\frac{T}{2} > T > 0$:

$$v_{Out} = \frac{V_{CC}}{2} - \frac{1}{R_1 C_2} v_{in} t + V'_{C2} \Big|_0^{\frac{T}{2}}$$

$$v_{Out} = \frac{V_{CC}}{2} - \frac{1}{R_1 C_2} v_m \frac{T}{2} + V'_{C2} \qquad (64)$$

Assuming $V'_{C2} = 0$, and $\frac{1}{R_1 C_2} \frac{T}{2} = 2$

$$v_{Out} = -\frac{V_{CC}}{2}$$

But, because there is no any negative supply, output will be clipped at 0 V

Thus $v_{Out} = 0\ V$ (see figure 40)

If R_1 is a constant value (fixing one value to calculate other):

$$C_2 = \frac{T}{4R_1} = \frac{1}{4R_1 f} \qquad (65)$$

Alternatively, if C_2 is the constant value:

$$R_1 = \frac{T}{4C_2} = \frac{1}{4C_2 f} \qquad (66)$$

For the next semicycle, negative semicycle:

Integrating along: $T > T > \frac{T}{2}$:

$$v_{Out} = \frac{V_{CC}}{2} + \frac{1}{R_1 C_2} v_{in} t + V'_{C2} \Big|_{\frac{T}{2}}^{T}$$

$$v_{Out} = \frac{V_{CC}}{2} + \frac{1}{R_1 C_2} V_m \frac{T}{2} + V'_{C2} \qquad (67)$$

In this stage: $V'_{C2} = -\frac{V_{CC}}{2}$, because at initial of this state when t=0, v_{out}= 0 V

Note that C_2 capacitor was charged from 0V to $-\frac{V_{CC}}{2}$ in the previous semicycle that why output $v_{out} = 0V$.

Continuing solving the equation (67) and keeping the same conditions for R_1 and C_2.

Now evaluating when t = T/2, $V_{out} = V_{CC}$

In the next semicycle (positive again):

Equation (64) is taken:

$$v_{Out} = \frac{V_{CC}}{2} - \frac{1}{R_1 C_2} V_m \frac{T}{2} + V'_{C2}$$

Where $V'_{C2} = \frac{V_{CC}}{2}$

Capacitor is charged at $\frac{V_{CC}}{2}$ at initial state and discharged to $-\frac{V_{CC}}{2}$. So, capacitor will be charging and discharging over cycles $\frac{V_{CC}}{2}$ and the output will be going to V_{CC} to 0 V respectively.

$$v_{Out} = 0\ V$$

For the next semicycle C_2 capacitor will be charged again to $\frac{V_{CC}}{2}$, and so on through the next cycles. So, output will be swimming between V_{CC} and 0 V. Note that swing of the signal is up and down from the DC reference voltage.

To ensure full swing in the output, V_m of the input signal must be:

$$V_m \leq \frac{V_{CC}}{2} \qquad (68)$$

Figure 40 is a plot of signals in the integrator circuit. Note that output is shifted by DC level.

The first semicycle is represented in the graph of the figure 40. The initial voltage of C_2 is assumed to 0 V.

Other signals can be also integrated. Triangle, sinusoidal, etc. But for these signals is necessary to adjust constants (R_1, C_2) to avoid saturation (clipping) or a very low amplitude.

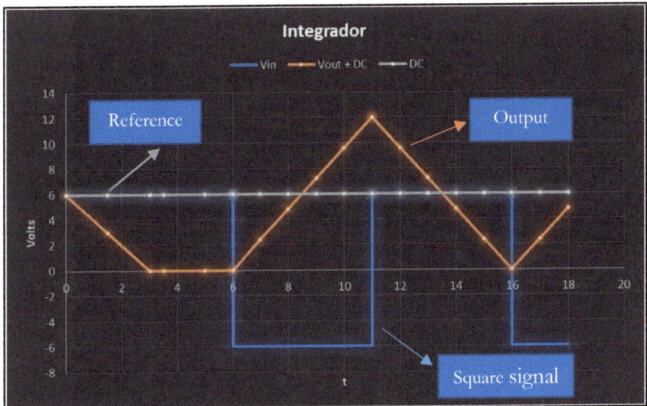

Figure 40. Plot of signals in the Integrator

In essential R_1 and or C_2 need to be adjusted or calculated every time frequency is changed. Otherwise the output amplitude will be varying along the frequency changes, even that the integration is accomplish for every input signal. This acts like a low pass filter transfer function.

Remember that the circuit is clipped on DC voltage so, the output will be always containing a DC level no matter if there is an input or not.

Next is the table 6 with the smart summary for this case.

✓ **Project:** Inverter Integrator

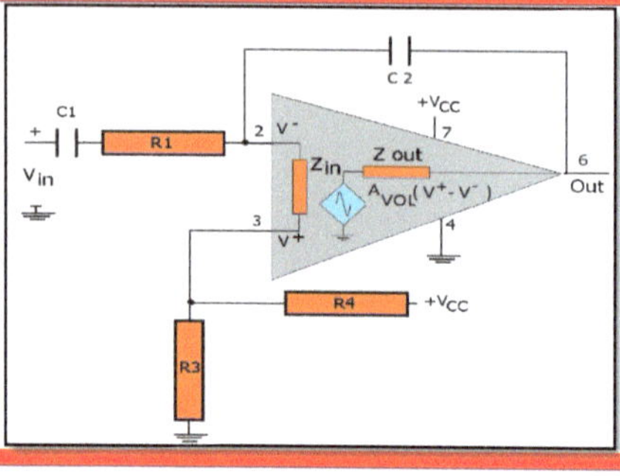

Parameter 1: ☐

$$V_{Out} = \frac{V_{CC}}{2} - \frac{1}{R_1 C_2} \int_0^t V_{in} dt + V'_{C2} \quad (63)$$

Parameter 2: ☐

$$V_{Out} = \frac{V_{CC}}{2} = \text{ with } V_{in} = 0$$

Parameter 3: ☐

$$C_2 = \frac{T}{4R_1} = \frac{1}{4R_1 f} \text{ ; square input}$$

Parameter 4: ☐

$$R_1 = \frac{T}{4C_2} = \frac{1}{4C_2 f} \text{ ; square input}$$

Parameter 5: ☐

$$V_m \leq \frac{V_{CC}}{2}$$

Comments:

Single supply

Table 6. Smart Summary of inverter Integrator

3.8 Charge Amplifier (photo detector)

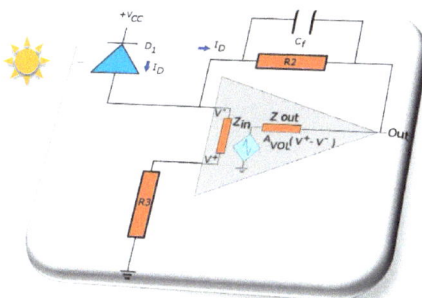

Figure 41. Charge amplifier.

Figure 41 shows a charge amplifier circuit. Diode D_1 can be a PIN photodiode like the BPW34. Diode is working here in the reverse zone, it means in transconductance mode. In this configuration diode D_1 is used to sense any light pulse coming from outdoor or a chamber if it is enclosure.

In dark operation, no light is incident on the sensitive face of the photodiode, the current in the diode will be only due to its impedance, and in reverse this impedance is very high. This current is known as the dark current, and it is around some few of nanoamps.

When a pulse of light strikes the sensitive face of photodiode then a photocurrent is produced. The amount of charge current produced is proportional to the intensity of the incoming light. But also, there is an internal process of photoconversion that fixes the efficiency between the energy of the wavelength incoming and the effective number of electrical charges produced in the semiconductor. So, for a fixed intensity the effective current is a function of the wavelength of the incident light.

In this manner, the output current of the diode will depend on the intensity and the wavelength associated to the incoming light.

For this example, let's suppose we want to detect a light pulse that are in the 100% efficiency region of the photodiode characteristic curve.

With the efficiency and number of photons incoming, the total output charge current can be easily calculated.

So, let's suppose now that diode is reverse biased, and the dark current is I_d

Supply voltage here is dual. But, can be also configured to work with single supply.

If not light is incoming, and the temperature is fixed:

$$V_{out} = -I_d R_1 \qquad (68)$$

At this moment the value of V_{out} represents the offset voltage due to the dark current. Notice it is negative, as it is an inverter amplifier.

So, to keep this voltage very low R_2 must be in the order of few megaohms as maximum, and the dark current in the range of nanoamps.

Notice that if photodiode is not blind at all, and some light is detecting, this level can be raised to even the negative saturation voltage. To avoid this, diode must be completed blind from any ambient light.

Now suppose a pulse of light is incident in the photodiode, and the form of this pulse will be:

$$i_{in} = Q \mid_0^t$$

Where Q is the charge collected in the photodiode in the time t.

The output will be:

$$v_{out} = -\frac{1}{C_f} \int_0^t Q + Vcf' \qquad (68)$$

Where the initial capacitor voltage is: Vcf'.

As a result, a negative pulse is also obtained in the output, but the charge current is being now converted into a voltage pulse with amplitude dependency of the relation:

$$\frac{Q}{C_f}$$

Pulse time is supposed to be shorter than $R_2 C_F$ constant. After the pulse is formed, capacitor is being discharged immediately through R_2.

The discharge of the capacitor will have the form:

$$v_{out} = -v_{out}\, e^{-t/\tau} + Vcf'(1 - e^{-t/\tau})$$

Where $\tau = R_2 C_f$

So, after some time pulse will be extinguished and the offset voltage will remain as product of the DC dark current.

Keep in mind that pulses wit short time are being considered here.

If light is a continuous function the output will be according to the equation (68) and negative saturation state may be reached.

The figure 42 shows a graphical solution of this circuit (pulse response)

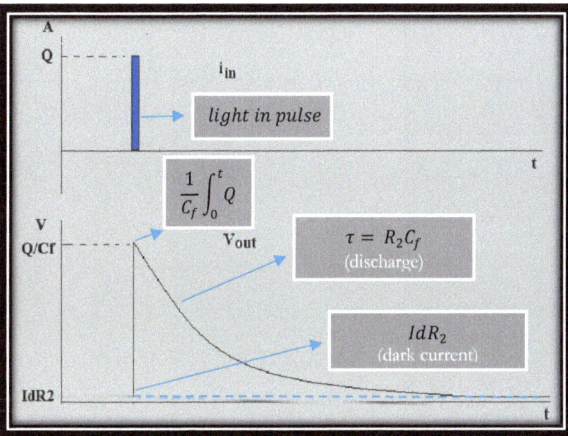

Figure 42. Pulse response of the charge amplifier. Note: output is inverted 180° to better view, but output is always negative.

Electronics: OP-AMP tech notes...! | 3.8 Charge Amplifier (photo detector)

✓ **Project:** Charge Amplifier

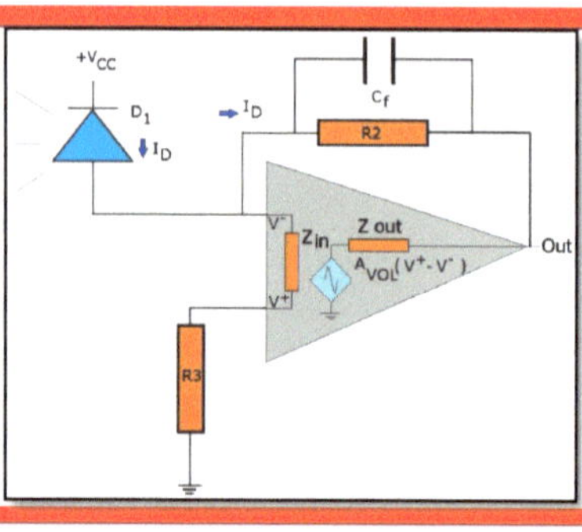

[Circuit diagram showing a photodiode D_1 connected to the inverting input of an op-amp with feedback capacitor C_f and resistor R_2, output labeled Out, with R_3 at non-inverting input to ground, powered by $+V_{CC}$]

Parameter 1: ☐

$$V_{out} = -\frac{1}{C_f} \int_0^t Q + Vcf'; pulse\ input \quad (68)$$

Parameter 2: ☐

$$V_{out} = -I_d R_1 \; ; continuous\ output.\ Dark\ current$$

Parameter 3: ☐

$$\tau = R_2 C_f;\ discharging\ time\ constant$$

Parameter 4: ☐

$$I_d \; ; dark\ current\ of\ photodiode$$

Parameter 5: ☐

$$R_2 \; ; in\ mega\ ohms;\ C_f \; ; in\ pF$$

Comments:

Dual supply

Table 7. Smart Summary of charge amplifier

3.10 Smith Trigger Comparator

Figure 43 shows the smith trigger comparator implemented with dual supply. But again, it can be configured to work with single supply. This is a convenient circuit for comparator applications due to its stability. The reference above and below zero ensure changes are made high enough from noise, avoiding undesired oscillations in the output.

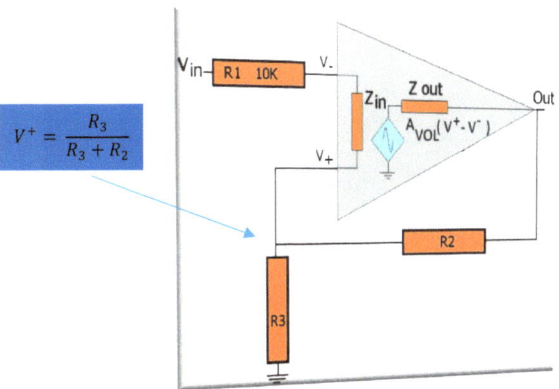

Figure 43. Smith trigger comparator.

Let's suppose again that: $V_{out} = +V_{oSat}$, and R_3, R_4 make a resistor divider.

$$V^+ = \frac{R_3}{R_3+R_2} V_{oSat} = V_{ref} \tag{69}$$

Then at this moment: $V_{in} > V^+$, the output changes immediately to: $-V_{oSat}$, thus:

$$V_{out} = -V_{oSat}$$

Then the reference will change also immediately to negative value:

$$V^+ = -\frac{R_3}{R_3+R_2} V_{oSat}$$

Assuming: $V_{oSat+} \sim V_{oSat-}$

Because reference has changed to a negative value, the output of the comparator will stay in the negative saturation voltage until the input in V⁻ port goes under the negative reference. At the moment it happens, output will change to positive saturation voltage, and the reference will be now changed to a positive value, so, output will remain in positive until input in V⁻ port goes again above the positive reference, and so on.

Changes in the output will occur when input crosses the reference voltage stablished in V^+ port.

Next, is the graphic that shows this behavior.

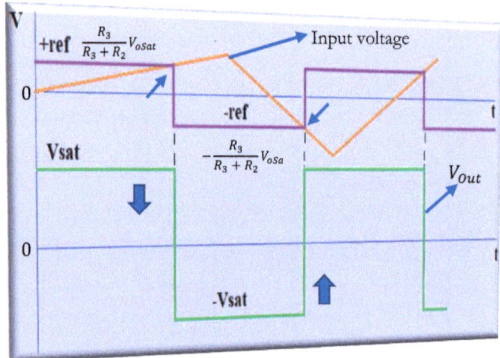

Figure 44. Smith trigger signals.

Saturation voltage will occur at some point near de supply voltage, most of time 1 to 1,5 V less.

So, for example if $V_{CC} = \pm15$ V, saturation voltage will be around ±13.5 - 14 V. Exception is made in the case rail-to-rail OP-AMP that reaches almost the supply voltage.

$$|V_{oSat}| \sim (V_{CC} - 1.5\ V)$$

This comparator has controlled hysteresis. The threshold hysteresis voltage can be controlled by changing the resistor divider form by R_2 and R_3 respectively. The positive and negative thresholds voltages correspond to:

$$V_{h+} = V_{ref}$$

And

$$V_{h-} = -V_{ref}$$

The total hysteresis voltage is: $2V_{ref} = V_h$

Concerning the parameters, the first thing to say here is the OP-AMP is working in a positive feedback loop. Positive feedback acts like a regenerator and tends to increase the output to saturation or make favorable to oscillations. The most important parameters here are: SR, and A_{VOL},

The SR parameter here is important because the output is a square form, so the edges of the output signal needs to be very fast. Then SR must be as high as possible.

The A_{VOL} parameter will decrease very fast with frequency dependence. In order the comparator has the maximum saturation output with minimum differential input the A_{VOL} must be higher as possible.

As example the LM311 may be a good choice for a voltage comparator, its A_{VOL} = 200.000.

Electronics: OP-AMP tech notes...! | *3.10 Smith Trigger Comparator*

✓ **Project:** Smith Trigger Comparator

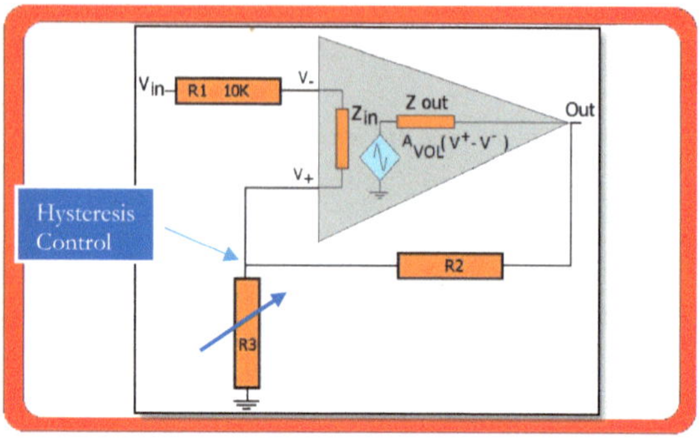

Parameter 1: ☐

$$V^+ = \frac{R_3}{R_3 + R_2} V_{CC} = V_{ref}$$

Parameter 2: ☐

$$|V_{oSat}| \sim (V_{CC} - 1.5\,V)$$

Parameter 3: ☐

$$V_{h+} = V_{ref}$$

Parameter 4: ☐

$$V_{h-} = -V_{ref}$$

Parameter 5: ☐

Comments:

Dual supply

Table 8. Smart Summary of smith trigger comparator.

Dr. Fernando J. Moutinho

4.0 OP-AMP Application table

Figure 45. OP-AMP: 8-pin assignment, package number J08A.

Model	Supply operation	Supply Voltage (V)	Zin (Ω)	Input bias (pA)	Input capacity (pF)	Noise e_N 1 kHz (nV)	Max. Load (kΩ)	GBP (MHZ)	Recommended application
LF351 TL081	Dual or single	±18 max. ± 5 min. 36 max. 10 min	10^{12}	50	-	25	2 1	4	Audio amp Charge amp Optical det Integrators Comparators Instrumentation
OPA27 OPA37	Dual or Single	±22 max. ± 4 min 44 max. 8 min.	$2\ 10^9$	15.000	-	4.5 (bajo)	1 1 2 1	8 OPA27 63 OPA37	Professional Audio amp Transducer amp Optical det Integrators Comparators Precision Instrumentation
CA3140	Dual or Single	±18 max. ± 2 min 36 max. 4 min.	$1.5\ 10^{12}$	10	4	40	1 1 2 1	4.5	Audio amp Charge amp Optical det Integrators Comparators Low power Instrumentation
LM111 LM211 LM311	Dual or Single	±18 max. ± 5 min 36 max. 5 min.	$1\ 10^4$ min	60.000 60.000 100.000	-	-	0.5 0.1 1 0.1	0.1	High speed comparators *(open collector Output) High power Driver Load (50 mA) TTL Compatible

Table 9. OP-AMP Application table.

Electronics: OP-AMP tech notes...! | OP-AMP Application table

Model	Supply operation	Supply Voltage (V)	Zin (Ω)	Input bias (pA)	Input capacity (pF)	Noise e_N 1 kHz (nV)	Max. Load (kΩ)	GBP (MHZ)	Recommended application
LF318	Dual or Single	±20 max. ± 5 min. 36 max. 10 min	$3\,10^6$	150.000	-	12	2 1	15 Fast SR = 50 v/μs	wide band amp A/D converter Oscillators Integrators Sample and hold Instrumentation
LF356	Dual Or Single	±18 max. ± 5 min 36 max. 10 min	$1\,10^{12}$	30	3	12	2 1	5 LF357 20	wide band amp A/D converter Optical det high speed Integrators Sample and hold Instrumentation
OPA341	Single (optimized) rail-to-rail	6 Max 2.5 min	$1\,10^{13}$	0.6	3	25	0.6	5.5 SR 6V/μs	Sensor biasing A/D converter Optical det Instrumentation Very low power
LT1006	Single (optimized)	20 max 2.7 min	0.25 10^{12}	10.000		22	0.6	1 SR 1V/μs	Amplifier A/D converter Optical det Instrumentation Very low power

Table 10. OP-AMP Application table. Continued

This page is intentionally left in blank

www.ingramcontent.com/pod-product-compliance
Lightning Source LLC
Chambersburg PA
CBHW041204180526
45172CB00006B/1185